D1357660

Wild, Wild World of Animals

# Domestic Descendants

A TIME-LIFE TELEVISION BOOK

*Editor:* Charles Osborne
*Associate Editors:* Bonnie Johnson, Joan Chambers
*Author:* Henry Moscow
*Writers:* Cecilia Waters, Deborah Heineman
*Literary Research:* Ellen Schachter
*Text Research:* Nancy Shuker
*Picture Editor:* Judith Greene
*Permissions and Production:* Cecilia Waters
*Designer:* Constance A.. Timm
*Art Assistant:* Patricia Byrne
*Copy Editor:* Eleanore W. Karsten
*Copy Staff:* Robert Braine, Florence Tarlow

WILD, WILD WORLD OF ANIMALS
TELEVISION PROGRAM
*Producers:* Jonathan Donald and Lothar Wolff
This Time-Life Television book is published by Time-Life Films, Inc.
Bruce L. Paisner, *President*
J. Nicoll Durrie, *Vice President*

THE AUTHOR

HENRY MOSCOW was formerly an editor of LIFE International and has written several books for TIME-LIFE BOOKS. Now a freelance writer, he has most recently coauthored *Educational Psychology: An Introduction* and is the author of *The Street Book: An Encyclopedia of Manhattan's Street Names and Their Origins.*

THE CONSULTANT

MICHAEL W. FOX, a veterinarian, psychologist and author, is currently Director of the Institute for the Study of Animal Problems, a division of The Humane Society of the United States, in Washington, D.C. Previously, he was Associate Professor of Psychology at Washington University in St. Louis, Missouri. The author of two popular books, *Understanding Your Dog* and *Understanding Your Cat,* Fox has also written several studies on wild canids based on his experiences in hand-raising timber wolves, coyotes, jackals and Arctic foxes.

THE COVER: Penetrating emerald eyes dominate the face of D'Artagnan, a three-and-a-half-year-old tiger-striped short-haired cat, another term for the very popular and common alley cat.

Wild, Wild World of Animals

# Domestic Descendants

Based on the television series
Wild, Wild World of Animals

Published by
TIME-LIFE FILMS

*The excerpt from "Toutouque," from* My Mother's House *by Sidonie-Gabrielle Colette, translated by Una Vicenzo Troubridge, copyright © 1953 by Farrar, Straus and Young (now Farrar, Straus & Giroux, Inc.), is reprinted with the permission of Farrar, Straus & Giroux, Inc., and Martin Secker & Warburg, Ltd.*

*The excerpt from* All Creatures Great and Small *by James Herriot, copyright ©1972 by James Herriot, is published with the permission of St. Martin's Press, Inc., and David Higham Associates, Ltd.*

*The excerpt from "The Big Gamble," from* Man o' War *by Walter Farley, copyright © 1962 by Walter Farley, is reprinted with the permission of Random House, Inc.*

*The excerpt from "The Story of Webster," from* Mulliner Nights *by P. G. Wodehouse, is reprinted by permission of the author's agents, Scott Meredith Literary Agency, Inc., 845 Third Ave., NY 10022.*

*"The Cat and the Moon" from* Collected Poems *by W. B. Yeats, copyright 1919 Macmillan Publishing Co., Inc., renewed 1947 by Bertha Georgie Yeats, is reprinted by permission of Macmillan Publishing Co., Inc., and A. P. Watts, Ltd.*



ISBN 0-913948-24-1.

Library of Congress Catalogue Card Number: 78-70986.

Printed in the United States of America.

# Contents

# Introduction

*by Henry Moscow*

IT IS VERY DIFFICULT TO IMAGINE HUMAN LIFE ON EARTH without domestic animals. For 12,000 years, *Homo sapiens* has made use of other creatures for food, transportation, labor, companionship and recreation. All the animals that man has domesticated were once wild, and the story of domestication is thus intertwined not only with the drama that unfolds in the history of civilization but also with the majestic, mysterious march of evolution.

With hindsight, these facts appear inevitable. Yet of the estimated three million species of creatures that share the earth with human beings, only a relatively small—though diverse—assortment has adapted to a way of life in association with them. On the other hand—fortunately for humans—there always seem to have been animals available that lent themselves to domestication and eased the task of survival, no matter how harsh or extreme the environment. Among such animals, domesticated first as a source of food and later for wool as well as for carrying burdens, is the yak of the Himalayas and Mongolia, a large, shaggy bovine that gives rich milk and thrives in thin mountain air and bitter cold. Another example, even more familiar to Westerners, is the reindeer—exploited since the end of the ice ages in Europe, at first only for its meat, hide and horns, but used much later as a beast of burden.

But far more ancient is the association of human beings with dogs, which are perhaps more familiar to more people than any other domestic animal today. Along with horses and cats, dogs are the exemplar of animals that are the friends and junior partners of human beings. Other creatures that contribute enormously to human material well-being, such as cattle, sheep and goats, have merely been subjected to man's dominance. This volume takes a look at animals in both categories that, like dogs and cattle, are more or less common in most of the world today.

In the process of adaptation from a wild state to domestic life, a few animals, including many dogs, have changed in shape, size, color and habits almost beyond any recognizable resemblance to their wild ancestors. A great many changes in domesticated animals, however slight, have been deliberately wrought by human beings in an effort to accelerate and manipulate the process of evolution and thereby encourage characteristics that make the animals more useful or more pleasing. The successes of such utilitarian breeding methods have been numerous, and an ever-increasing understanding of genetics promises more of them. For now, however, the furtherance of domestication depends mainly on three practices:

- Genetic selection for increased productivity, as of milk in cows and eggs in chickens, and specific attributes such as speed in racehorses, scenting capacity in bloodhounds and color in mink.
- Early exposure to human beings: Even puppies, offspring of the longest-domesticated animal, will become feral if they are isolated from people during their first 12 weeks of life.
- Imposition of physical changes, such as the castration of cattle and horses

*A detail from a 9th Century B.C. Assyrian relief depicts servants of King Assurbanipal hunting a lion with nets and a pack of fierce-looking mastiffs.*

to keep them calm, the debeaking of chickens raised in close quarters to prevent them from injuring one another, and the placing of rings in pigs' snouts to frustrate them from rooting among crops.

Some of these principles have been known since about 8000 to 6000 B.C., when agriculture began to replace hunting and people were settling down in permanent communities after a long nomadic existence. But though domestication may be recent in human history, its origins are largely mysterious; archeological evidence offers clues, but for the most part the experts can only speculate about the origins of the relationships between human beings and the animals that serve them.

A starting point for explaining these relationships is the fact that the wild ancestors of many of the most familiar domesticated animals were social, living in herds, small bands or even pairs of their own species. Since people are social too, they were also accustomed to the presence of other creatures of their species—the family group, for instance. Thus in their gregariousness, animals and men were predisposed psychologically to investigate each other. Some animals approached human encampments to scavenge, picking up a bone or leftover scraps from a kill:

Goats and pigs might have ventured near so often that their presence became customary and was tolerated because they cleaned up a campsite. But a permanent relationship probably was formed when human beings adopted young animals for a very human reason—a need for play and entertainment. In earliest times, an animal might have been released or eaten when it was almost full grown and less playful. But a pattern was gradually established that eventually led to the breeding of the more docile creatures and the raising of animals that never would know a life entirely without human supervision.

The first domesticated animals were dogs, and their ancestry has yet to be positively established. Some investigators favor the Asian wolf. It was small and, like all wolves, it was social. But recently ethologists have come to believe that the ancestor of the dog must have been a canid that is now extinct, and they theorize that it may have resembled the present-day Australian dingo, which once ranged all over Eurasia. Whatever this wild canid looked like, it probably first approached people in order to scavenge some of their leavings. Pups may have been adopted and in turn adopted humans as their pack, much as dogs do today. At some point, presumably, the usefulness of dogs was noticed and appreciated: Some of their senses were far more acute than early man's, and they could smell and hear danger long before he did. In addition, human beings had highly developed brains, which meant that they needed longer periods of deep sleep than other animals did. The pup that woke a sleeping family when it sensed the approach of a predator would certainly earn gratitude and affection. The same keen senses that warned of a predator's approach also functioned well in the hunt. With the advantage of their weapons, hunters may have learned to gain another edge—the cooperation of canine companions—by following dogs to the prey, clubbing it and later sharing morsels with the dogs. Working together, men and dogs came to control and guide, rather than merely pursue, animal herds. And gradually, sometime about 9000 B.C., herding began to share equal time with hunting.

*A Scandinavian rock carving dating between 1500 and 1000 B.C. shows a ship carrying a cargo of cattle. Found at the Swedish coastal town of Tanum, this and similar carvings were probably thought to have magical properties.*

Whether domesticated dogs originated in one region and spread out or whether they developed in a number of places independently remains undetermined. Apart from an excavation in Siberia that indicates that men and doglike animals worked in partnership more than 10,000 years ago, fossils also indicate that dogs living at ease with human beings ranged from Scotland and Denmark through Germany and Switzerland to the Middle East during the Neolithic Age, which began about 9000 B.C.; remains from about 8000 B.C. found in Switzerland demonstrate that even then dogs differed in size and shape from one another. Some of these dogs were very small and may have lived in primitive dwellings on the same sites—which would suggest that the tradition of the house dog is very ancient indeed.

The first animals to be herded, with or without the aid of dogs, were goats and sheep. Their association with human beings also probably predates agriculture—both animals were capable of being driven for long distances by nomads. Goats were almost certainly domesticated from the wild bezoar goat of present-day Iran, Israel and Jordan no later than 7500 B.C., and on the evidence from one archeological site near Bethlehem, possibly as early as 10,000 B.C. They provided meat and milk, and with their ability to exist on scrub vegetation, they could be herded in semiarid regions. Eventually, these hardy beasts were welcomed in much of West-

# Ancestral Models

*The animals—drawn to scale—that are illustrated on these pages are the closest living relatives of the wild species from which the domesticated animals in this book are descended. Some, like the mouflon, the wild pig, the wild goat, the African wildcat and the red jungle fowl, still exist in the wild and lead lives very similar to those led by their wild forebears, despite the drastic reduction of their natural habitat. Along with the domestic dog, the Asiatic wolf and the Australian dingo are thought to be descendants of a now-extinct common ancestor that ranged much farther across Eurasia than any of these survivors. Przewalski's horse no longer exists in the wild, but it has been preserved by conservationists in game parks and zoos. The aurochs represents a unique scientific experiment; it is the result of years of genetic back-breeding of strains of modern-day cattle to bring out characteristics similar to traits attributable to a wild ancestor. The resulting animal, which is five and a half feet high at the shoulder, is smaller than the ancestral form, but resembles it closely in temperament and build.*

Wild goat *(Capra aegagrus)*

African wildcat *(Felis libyca)*

Przewalski's horse *(Equus przewalski)*

Wild pig *(Sus scrofa)*

Aurochs *(Bos primigenius)*

Red jungle fowl *(Gallus gallus)*

Dingo *(Canis dingo)*

Asiatic wolf *(Canis lupus)*

Mouflon *(Ovis musimon)*

ern Asia, North Africa and all over Europe. The wild ancestor of domesticated sheep, the mouflon, also originated in Asia and in some regions may have been domesticated earlier than the goat. When human beings first began herding sheep near settlements, they raised them for meat—which was tastier and tenderer than that of goats—and only later for milk. The animals' coats, however, were hairy rather than woolly, and it was 7,000 years before crossbreeding produced the luxurious wool that was a prize staple of the Romans.

More highly productive of flesh and fat than goats or sheep, wild pigs, in a variety of sizes, ranged everywhere in the Old World from China through Mesopotamia (roughly present-day Iraq) and Europe. But scavenging pigs were not domesticated until people had begun cultivating crops. Pigs, though intelligent and affectionate, are not tractable enough to be herded over great distances, as nomadic peoples would have required. The Chinese have been keeping and eating pigs for at least 6,000 years, and an ivory figure, from about 2700 B.C., of an obviously domesticated pig—fat and without bristles—has been found at a site of Sumerian civilization in southern Mesopotamia.

Even earlier than this, however, the Egyptians were employing pigs as laborers. The Egyptians loosed pigs in their fields, where the porkers, grubbing energetically for roots and worms, broke up the clods of earth; in addition, their feet made holes just the right size for planting. But the Egyptians were ambivalent about eating the meat of swine, and at various times taboos proscribed pork entirely or restricted its consumption to certain days of the week. Like other animals in Egypt, the pig was sacred to a god, but the pig's was the evil deity Seth. The Jewish and Islamic aversion to pigs may have derived from the Egyptian; its basis may have been the reputation of pigs for wallowing in and eating filth. But a more practical explanation may be that an enormous increase in Middle Eastern populations led to deforestation. The forest floor, where the pigs found ample forage, was replaced by fields, with encroaching grassland and desert. The pig, unlike goats and sheep, began to compete with man for the available grain crops, and pig raising became uneconomical. Certainly the prejudice never extended to Europe, where natural forage abounded, and farmers valued the prolific and easy-to-care-for pig.

Nobody knows just when men began domesticating cattle, historically more prestigious than pigs. Drawings that are at least 12,000 years old on a cave floor of the Grotte de la Mairie in France's Dordogne region depict quiescent cattle that might possibly have been under man's dominance, but the pictures omit halters, which would be confirming evidence. However, remains dating from 6500 B.C., in Neolithic times, found at the sites of prehistoric communities along the lakes of Switzerland, are certainly those of domesticated cattle. They were aurochs, the ancestors of the domestic cattle common to America, Europe and Asia. Aurochs were a species of large wild cattle believed to have originated in northern India and migrated over most of Europe, to North Africa, Asia Minor and as far east as the shores of the China Sea. For millennia, the beasts, as awesome in their size as bison, were hunted.

As men made their first attempts at agriculture, aurochs probably ventured among the crops to forage. The farmers would have reacted by driving the herds off; zöologists speculate that in their flight the cattle left behind newborn calves. When the calves were discovered, it is likely that people took them close to their settlements and fenced them in or tied them up. It is also possible that the farmers captured pregnant cows too heavy to run away. This nucleus of a herd would increase when wild bulls sought out the cows at rutting season and were allowed to sire young—calves that would grow up accustomed to husbandry.

Where and when the daredevil lived who first attempted to ride a horse remains an unanswered question. The oldest known evidence of a man on horseback is a wood sculpture found in an Egyptian tomb dated 1350 B.C. But for thousands of years horses were hunted for their meat. Frequently herds were driven over a cliff; the bones of some 10,000 horses lie at one such site in France. Then about 3500 B.C. the people who lived in what is now the Ukraine region of the U.S.S.R began to keep horses, and the first rider may well have been a youth who raised a newborn foal to maturity. The horses of the Ukraine probably were of the kind now known as tarpans; wild herds of these horses survived into the 19th Century despite relentless hunting by peasants.

It was the Scythians, the fierce, wealthy nomads who roamed the Eurasian steppes with their herds of sheep and cattle, who first truly mastered horsemanship. They rode only geldings, keeping the noblest stallions in stud for selective breeding. And it was the Scythians in their wide-ranging military forays who demonstrated to the Egyptians and the Mesopotamians the superiority in battle that a good horse bestowed on a fighting man. The Hyksos people, who attacked Egypt from the east between 1780 and 1710 B.C., demonstrated another warlike use of horses: They invaded with horse-drawn chariots, a military innovation as formidable in its earliest days as the tank was in World War I, and they crushed the Egyptian resistance. Within the next 500 years, horse-drawn chariots and cavalry became the most important elements of armies. Reverence for horses continued into the Christian era, and by the Middle Ages the saddle horse, symbolized by the war charger of the armored knight, was the most potent of status symbols.

Smaller relatives of horses, lowly asses also once enjoyed high esteem. First domesticated in Egypt or Libya, possibly centuries before 1650 B.C.—the date of an Egyptian panel depicting them among other domestic animals—asses were prized by the Egyptians and the Romans as well. But their two greatest virtues—stolidity and an ability to live on very little—eventually failed to impress their wealthier masters, who came to prefer showier creatures such as horses. By the time asses reached Europe in the early Middle Ages, they had become poor men's steeds and beasts of burden.

Like asses, cats have known both good and bad times, but the bad times were horrendous. Domestic cats descend from the African wildcat, or Libyan cat, and their first known contact with human culture appears to have been sometime before 2500 B.C. in Egypt. As wild cats they hunted the rats that fattened themselves in Egyptian granaries. In growing admiration, the Egyptians promoted cats from this humble but useful status and trained their feline pets to fish and to hunt birds. Eventually, the Egyptians elevated cats to the level of gods. When Christianity, rising and spreading, encountered cats, which the Romans had introduced all over their empire as far west as Britain, adherents of the new religion abhorred the animal's pagan divinity and equated it with evil. In the name of religion, cats were

killed; even the 15th Century pope Innocent VIII encouraged their destruction.

But cats survived—demonstrating their legendary possession of nine lives—and their utility as rat and mouse catchers made them indispensable to society. The appeal of their grace and beauty also saved them, and in the last few centuries cats have been generally free from persecution. Some would say that cats have never been truly domesticated; in fact the animal psychologist Michael Fox suggests that house cats may have domesticated people, by managing to get their human owners to do what they want instead of the other way around.

In contrast, domestication of the chicken was easy. The hen busily laying 250 eggs a year in a modern food factory is descended from the wild red jungle fowl, a hardy, flying Asian pheasant. The red jungle fowl normally deposits 30 eggs a year, but if her nest is robbed repeatedly she will lay as many as 80. Someone discovered that systematically removing eggs from the nest stimulated further laying, and this simple procedure culminated, around 2000 B.C. in India, in the earliest domestication of the birds; by the Sixth Century B.C., chickens were the dominant domesticated fowl all over Asia and around the Mediterranean.

Difficult though it may be to visualize modern life without domestic animals, the

*Two cows—one of them being milked—and a calf dominate the center of this panel, a replica of the west side of the tomb of Ashayt, a favorite of the Egyptian king Mentuhotep, who reigned about 2000 B.C. Ashayt is seated at each end of the panel; flanking symbols represent the abundance of food and drink that will accompany her, along with the cattle, into the afterlife.*

human species has survived for more than 90 per cent of its total history with no dependence on them. It is indisputable, however, that civilization could scarcely have evolved in their absence. The creatures that societies transformed changed societies in their turn; as animals were taken into the service of infant civilizations, they supplemented the cultivation of crops and nurtured the continuance and progress of a human culture. If domestic animals had provided no more than a dependable food supply, that alone would have made it possible for human beings to explore the globe and begin to probe the universe.

It is conceivable that technological societies may embark on a future life in space without the services of animals; for the present, however, as populations increase and the earth's resources are threatened and shrinking, the multitudes of breeds that human beings continue to create may not be equal to the task of ensuring man's survival. Experts caution that both the wild ancestors and hardy early breeds must be preserved as a reliable genetic resource that has been tested over the centuries. As for the animals themselves, domestication so far has saved them from the extinction that has been the fate of many wild species; it has been, as one scientist has called it, "the greatest conservation movement of all times."

# Dogs

No mammal species today is more diverse in shape, size, color and character than the domestic dog. Except for man, none is more versatile nor occupies so many different habitats. Some 400 breeds of *Canis familiaris* exist, and the number is growing steadily as humans continue to develop breeds with ever-greater potential as champion show dogs, sporting dogs, working dogs and house dogs.

Bred over millennia for docility, probably no other animal is so dependent on human beings. But despite the dog's intimacy with man, it enjoys a radically different sensory world. The dog is able to hear sounds having a frequency of 30,000 cycles per second, far above the human limit of 18,000. It also has a sense of smell incalculably superior to man's. Besides its sensitive nose, the dog has additional olfactory receptors in the roof of its mouth. And some scientists claim to have found infrared receptors in the dog's nose that react to infinitely tiny amounts of warmth, which explains the ability of the St. Bernard to locate live people under seven feet of snow—but not those that are dead. Compared to its sense of smell, a dog's eyesight is generally poor, and the animal is color-blind.

However, the dog may have complex, undiscovered sensory abilities that allow it to receive more signals from its surroundings than are now recognized. Such capacities would help explain the dog's extraordinary ability to find its way home and even to orient itself in totally unfamiliar circumstances. Among numerous strange cases of the latter phenomenon, the story of Prince, a mongrel, is a classic. During World War I, Prince mysteriously found his way across the English Channel and joined his master in the trenches in France.

By definition, the working dog best exemplifies how canine potentialities are most usefully combined with human skills to develop the dog's temperamental and sensory strong points. Its traditional roles are well known: herding sheep; hauling sleds in the Arctic; and pointing, flushing and retrieving game. Today's working dogs also serve as lifeguards and as soldiers detecting mines and going out on patrols. Other specialists guard commercial establishments, guide the blind, aid therapists working with the emotionally disturbed, and operate as police detectives. In one recent year, the canines of Scotland Yard, which counts 280 dogs—mostly German shepherds—in its ranks, participated in 7,414 arrests, helped find 65 missing persons, and helped recover 224 items of lost property.

The show dog, bred to the exacting standards of kennel clubs, excels in the looks and demeanor considered proper for its breed. But unlike natural selection, a slow process that produces adaptive changes over the course of eons, artificial selection is rapid, and professional breeding can effect changes within just a few generations. Selection for show-ring esthetics alone may submerge once-inherent abilities. Water rescue, for instance, is instinctive in Newfoundlands. At present, however, at least one breeder has found it necessary to establish a swimming school for the dogs to reinforce the breed's aquatic instincts.

A sociable temperament is probably a dog's most important attribute: Most people are concerned chiefly with the house dog, or pet, in its major role as companion. A house dog often considers itself part of a human family. However, this attitude is not entirely instinctive; in order to form a satisfactory attachment to its owner, the dog must be socialized as a puppy, becoming accustomed to human contact before the age of 12 weeks.

Few house dogs, of course, would completely ignore fellow canines even if they could. Like many other animal species, the dog is concerned with its social status, or place in the pecking order. The dog has a keen sense of territoriality and stakes out its turf by urinating. It also respects the territorial rights of others: A 16-ounce Chihuahua on home ground can bark off a 200-pound Great Dane. On neutral territory the dog urinates to inform other dogs that it has been there; when it meets another dog, they determine quickly, after a few sniffs and threat displays, which outranks which. The social inferior may then roll over—one way that dogs indicate submission.

Lacking contact with other dogs, a dog's life can be just that, and in the saddest sense. House dogs that are excessively dependent on human beings may develop emotional problems and psychosomatic illnesses. For instance, a dog's status as a cosseted favorite may suffer a severe blow when an infant is introduced into the family, and the animal may develop symptoms such as asthma, compulsive eating, epilepsy or paralysis. The house dog's sexual needs are often left cruelly unsatisfied; when other solutions are impractical, castration or spaying is a kindness. Companionship, affection and firm guidance are important for a house dog in keeping it contented, but so is the chance to discover—and relish—the fact that it is a dog, not a substitute human being.



*Mastiff*

# Sporting Spaniels

Among the earliest breeds of dog, the spaniel may also have been one of the first to be bred and trained for sophisticated and specialized contributions to the hunt. Perhaps of Spanish origin, as its name suggests, the spaniel was specialized with the development of land and water varieties as early as the 14th Century. In succeeding decades the breed was categorized by size and by proficiency with different types of game. Water spaniels were used in duck hunting, and land spaniels hunted upland game birds, tracking them by scenting the air, then springing toward them to flush them into visibility—and often into a net. A modern spaniel used as a gun dog briskly quarters in front of a hunter, crossing the field laterally while moving constantly ahead. After flushing a bird from its cover, the dog remains motionless as the gun is fired, and often it follows up by retrieving on command.

The smallest of the sporting spaniels, the lightweight cocker is so called because it once specialized in the pursuit of woodcock. However, the American version of the cocker became irresistible to pet owners, and indiscriminate breeding to satisfy the demand diminished its hunting instincts—its body structure and benign temperament suffered as well. By the end of the last century, the Brittany spaniel of France had also been excessively inbred. But one French sportsman took it upon himself to restore the Brittany's quality, and by patiently and intelligently introducing crosses of other breeds, he brought the dog back to its former sporting prominence.

*An English springer spaniel (left) gently holds a retrieved grouse in its soft mouth. A favorite ally of the sport hunter, the agile springer with its dense coat is at home in brambles or swamps. It is particularly fine at flushing pheasants, and it is also a willing swimmer that is adept at retrieving ducks.*

*Hardy and dependable, the Brittany spaniel (right) is also sensitive and affectionate. Closely resembling a setter, it both points and retrieves game for the hunter.*

*An American cocker spaniel (below) sprints across a field with the ebullience that made it the most popular pet-dog breed in the United States from 1946 to 1952. If well trained, some cockers still make good hunting companions.*

*Head held high, an English setter points its quarry. This rugged breed is more than 400 years old and was probably the first purebred hunting dog employed in the American colonies. One English setter became so famous at field trials in America that after its death it was stuffed, mounted and displayed at the Carnegie Museum in Pittsburgh.*

*A Gordon setter (left) races eagerly toward a fallen bird. Gordons attained their popularity as hunting dogs in the kennels of the Scottish Duke of Gordon in the late 18th Century. Intensely loyal, the sturdy black-and-tan Gordons are prized as pets and as hunters. Like all the leggy setters, they need a great deal of exercise.*

*A duo of lithe Irish setters (right) frolic on a beach. Set apart as a breed by their solid dark-red color, they are noted for their gentleness and the gaiety of their dispositions.*

# Long-legged Roamers

The British, with a traditional passion for both sport and animals, are responsible for developing many of the sporting breeds of dogs, among them the setter. The setter was mentioned in English writings as early as the 16th Century. Once called "a spaniel improved," the dog's equipment includes long limbs, great speed and endurance over distances, a deep chest with expanded lung capacity, a long silky coat for beauty as well as protection against the cold, and an extremely keen nose.

The setter's name comes from an early hunting role. Four hundred years ago, game birds were often captured by hunters casting a net over them, and it was necessary for a hunting dog to locate the game and not only stay motionless at the moment of capture but also keep out of the way. Hence dogs were trained to "set," or crouch low to the ground, and some became known as setters. Breeding for responsiveness, sociability and eagerness to please—mandatory qualities in a hunting dog—resulted in setters that were devoted and attractive companions.

The vivacious Irish setter first won acclaim in 18th Century Ireland, where it is usually a mixed red-and-white color. By the 1920s in America, however, it was considered a fault if the setter was not solidly red. Perhaps because its beauty made it so decorative, its hunting instincts were often ignored, but its tendency as a pet to roam is a reminder of past skills in the field.

# Wedded to an Instinct

A well-bred pointer pup with no previous hunting experience shows an acute interest when confronted with small game. Without being prompted, the pup may instinctively demonstrate its ability to point—to take an immobile stance and stare at the prey intently, an action that in the field indicates to the hunter the presence and position of a game bird.

Although some breeds of hunting dog adapt quite well to life as house pets—if they can get enough exercise—it is probably very unfair to keep a pointer simply as a pet without employing its highly developed instincts. In fact, a pointer not used as a hunter may take to chasing substitute and possibly dangerous "prey" objects such as automobiles. The pointer is so wedded to its innate talents that one tall tale popular in Victorian England is almost believable. It was claimed that a hunter lost his pointer on the moors and a year later discovered the skeleton of the dog, still pointing at the skeleton of a bird.

*An English pointer (left, top) shows fine form in its pointing stance. One of the earliest English breeds, the pointer dates from the mid-16th Century, when it was used to spot hares for coursing greyhounds. In the 19th Century, the breed was crossed with even-tempered setters to produce pointers with a relatively docile disposition.*

*A German short-haired pointer (right) awaits a command. The breed adapts readily to water, and the dogs are also reliable upland game birders. More heavily boned than English pointers, they are not as swift*

*A wire-haired pointing griffon (left, bottom) is ready to romp. A good retriever, the pointing griffon also tends to be high-strung. This breed, with its short bristly coat, is the result of the work of one 19th Century Dutch breeder who spent 12 years developing the dog in Germany and France.*

# Gutsy Gun Dogs

Playful dogs have always delighted in racing after sticks and bringing them back to their owners. Retrievers such as the Labrador and the golden, however, carry this instinct to a high degree of professionalism. The development of these two breeds probably was a response to the use of the modern long-range rifle in the 19th Century: Hunters needed a hardy, courageous dog willing to run swiftly over long distances in the roughest terrain and lead them to wounded birds or retrieve downed prey.

Especially bred for hunting ducks and for endurance in the coldest climates, neither the Labrador nor the golden flinches from entering the iciest pond to retrieve a fallen bird. The Labrador is an accomplished swimmer—water is said to run off its dense coat like oil. It is a mixture of water spaniel, setter and St. John Newfoundland (a favorite of fishermen that was brought to England about 1835) and it was not recognized as a distinct breed until 1903. The golden retriever has much the same heritage, but one of its ancestors was a yellow pup in a litter of wavy-coated black retrievers. The oddball pup so intrigued a late-19th Century English lord that he began breeding for a fine hunting dog with a handsome golden coat.

*The Labrador retriever (right), guarding the day's bag of ducks, is misnamed: One of its ancestors was from Newfoundland. A swift swimmer, the Labrador has a flat "otter tail" that it uses as a rudder.*

*Paddling steadily, a golden retriever brings a hunter's quarry to shore. At 70 pounds, the golden is muscular and powerful as well as obedient. Quick to accept training, it is often used as a guide dog for the sightless.*

*Nose to the ground, a young bloodhound (left) picks up a scent. The bloodhound has the best nose in the tracking business and serves as a professional tracer of lost or errant persons. Its amiable, docile disposition belies it sanguinary name.*

*A Basset's head (right) resembles that of the bloodhound, one of its progenitors. But the Basset's legs are much shorter, and the dog is more heavily boned for its size than any other breed. Its keen nose is second only to the bloodhound's, and it is genial despite its forlorn face.*

*A beagle (left, bottom) peers out of its house at the snow. A small hound, the beagle adapts to almost any climate. In pairs or packs the lively dogs are excellent rabbit hunters.*

*One of the ancestors of the dachshund (below) tracked the ferocious badger; dachshunds today hunt rabbits.*

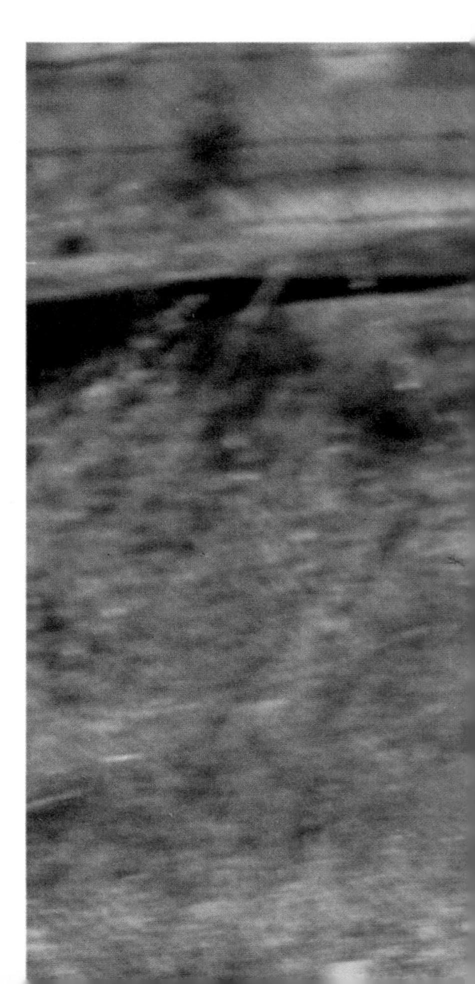

# The Nose Knows

A hound's capacity for picking up a scent and following it reaches uncanny heights in the bloodhound, an ancient breed that goes back to pre-Christian times. Stories of the bloodhound's acuity abound. A Kentucky bloodhound named Nick Carter successfully followed a trail that was more than 100 hours old; another dog tracked a criminal across 135 miles in Kansas; yet another rounded up 23 escaped convicts in a day and a half. For that matter, the dog's association with crime is ancient indeed. A 1527 English law stated that any person denying a trailing "sleuth hound" entrance became an accessory to the crime. Early English law also declared that a bloodhound ending up on one's doorstep was as good as a search warrant. The bloodhound's "testimony" is admissible evidence even in most American courts. But this well-mannered dog hasn't a vicious bone in its 85-pound body; it follows a trail for the sheer canine pleasure of the pursuit.

Beagles, Bassets and dachshunds also have keen noses. The beagle, one of the oldest breeds, closely resembles the original hound. The Basset can be traced to monastery dogs of the Middle Ages, and an oddly dachshund-like dog even appears in a statue of an Egyptian pharaoh.

# Barks and Bites

With its acute senses and devotion to duty, the German shepherd easily made the transition early in this century from the stock guarding its name implies to police and war work. Then after World War I its popularity with the general public soared, and the breed's temperament degenerated as a consequence. Today, however, many would confirm that a well-bred, well-trained German shepherd is one of the canine kingdom's finest guard dogs.

The aim of the Doberman pinscher's breeder, one dog historian claims, was to produce a giant terrier that would attack anything. Recent crosses have gentled the breed, but it still may be true that for a person with a Doberman as a friend, enemies are scarce. A much older breed is the English mastiff, whose tactics with malicious strangers were noted in 1792 by the famed naturalist Linnaeus: The dog would "seize the person, throw him down, and hold him there for hours, or until relieved, without biting."

The world's first early-warning system may have been the barking dog. In any case the alert Lhasa apso has been operational for 800 years in the monasteries and villages of Tibet, where it is called the Bark Lion Sentinel Dog.

*Intimidation by sheer size is the forte of the mastiff (left), which can weigh as much as 210 pounds. Its instincts lead it to be highly protective of and devoted to its owners, and it is gentle enough to be the guardian of children*

*The fearless and speedy Doberman pinscher (right) is named after its originator, a German named Louis Dobermann, who developed the breed in the 1890s. In the United States, the dog's ears are usually cropped to stand erect, but in Britain, where this dog was photographed, that is illegal.*

*A German shepherd (below) is a loyal companion as well as an intrepid guard dog. Dignified and sensitive, shepherds respond better to praise from their owners than to scolding. The shepherd has remained a natural-looking dog, unchanged by show-ring breeders' fancies.*

*Scarcely a sight to strike fear into the heart, the 15-pound Lhasa apso (above) is, however, a superb watchdog of long standing.*

# My Mother's House

## by Sidonie-Gabrielle Colette

*Born in France in 1873, Sidonie-Gabrielle Colette was the author of 15 novels, all of which were characterized by her intimate style and her sensitive insights into the emotions of her characters. In one of her later works,* My Mother's House, *Colette tells of her own childhood, her happy home and her mother's menagerie of domestic pets. In the excerpt below, Colette tells a story of love and jealousy revolving around Toutouque, a charming bulldog.*

Broad and squat as a four months' pigling, smooth-coated and yellow with a black mask, she looked more like a diminutive mastiff than a bulldog. Vandals had trimmed her shell-like ears into points and had cut her tail close to her behind. But no dog or woman in the world ever received, as her share of beauty, eyes that could be compared to those of Toutouque. When my elder brother, then a volunteer in the neighbouring county town, rescued her, by bringing her home to us, from an idiotic order condemning to death all the barrack dogs, she bestowed upon us one glance of those eyes the colour of old Madeira, a glance hardly at all anxious, but full of perception and shining with a moisture like that of human tears. We were all conquered on the spot and Toutouque was allotted her ample place before the wood fire. Everyone—and especially I who was then a little girl—felt the charm of her nanny-like amiability, and her equable temper. She barked very seldom, a muffled and throaty bark, but she talked a great deal in other ways, expressing her opinion with a flashing smile of black lips and white teeth, slyly lowering her dusky lids over her mulatto woman's eyes. She learned our names, a hundred new words and the names of all the she-cats, as quickly as an intelligent child. She took us all to her heart, followed my mother to the butcher's, and accompanied me daily a part of the way to school. But she belonged only to my elder brother who had saved her from the bullet or the rope. She loved him with an intensity of love that made her self-conscious in his presence. For his sake she became foolish, bowing her head and positively asking for the torments that she waited for as rewards. She would lie on her back, exposing her belly studded with purplish teats upon which my brother would strum, pinching each in turn, the tune of Boccherini's *Minuet.* The rite demanded that at each pinch Toutouque—who never failed to respond—should utter a little yelp, at which my brother would exclaim severely: "Toutouque, you're singing out of tune! Begin again!" There was no cruelty involved, the merest touch drew from the ticklish Toutouque a series of varied and musical cries. The game ended, she would remain supine, asking for more!

My brother warmly reciprocated her devotion, and composed for her benefit those songs which burst from us in moments of unchecked childishness, queer jingles born of rhythm and repetitions of words blossoming in the innocent vacancies of the mind. One refrain lauded Toutouque for being

*"Oh, yellow, yellow, yellow.*
*Inordinately yellow,*
*Oh, uttermost extremity of yellow. . . ."*

Another song celebrated her massive build and named her three times over, "endearing cylinder," to the lively cadence of a military march. Then Toutouque would roar with laughter; in other words, she would bare every tooth in her nubbly jaw, lay back what remained of her clipped ears, and in default of a tail to wag would wag her weighty posterior. Whether asleep in the garden or gravely occupied in the kitchen, the Song of the Cylinder, chanted by my brother, would bring Toutouque straight to his feet, captivated by the familiar strains.

One day when Toutouque lay roasting herself, after dinner, upon the burning marble of the hearth, my brother, seated at the piano, incorporated the Song of the Cylinder, without words, in the overture he was reading at sight. The opening bars hovered over the animal's slumber like importunate flies. Her coat, smooth as that of a Jersey cow's, twitched here and there, and her ears . . . An energetic repetition—piano solo—half opened the eyes, full of human confusion, of the musical Toutouque, who rose to her feet and asked me clearly: "Haven't I heard that tune somewhere?" Then she turned to her ingenious tormentor who was persistently hammering out the favourite air, accepted from his hands this new magic, and went and sat close to the piano in order to listen better, with the knowing yet mystified expression of a child trying to follow a conversation between grown up people.

Her gentleness was a rebuke to all teasing. She was given new-born kittens to lick and the puppies of stranger bitches. She kissed the hands of toddling infants and allowed young chickens to peck her. I was inclined to despise her for her over-fed convivial meekness until the day came when Toutouque, in due season, lost her heart to a gundog, a setter belonging to the local restaurant keeper. He was a big setter, endowed as are all setters with a "Second Empire" charm; a red-blond, long-haired and with luminous eyes, he lacked character but not distinction. His mate resembled him as a sister, but was nervous and subject to vapours. She uttered shrieks if a door was banged and wailed at the sound of the Angelus. For purely euphonic reasons their master had named them Black and Bianca.

This brief idyll brought me to a fuller knowledge of our Toutouque. Walking with her past the cafe, I saw the red-haired Bianca lying upon its threshold, her paws crossed, her uncurled ringlets curtaining her cheeks. The two bitches exchanged a single glance, and Bianca fled behind the bar shrieking the shriek of the crushed paw.

Toutouque had not stirred from my side, and her tipsy sentimental eye enquired with astonishment: "What can be the matter with her?"

"Let her alone," I replied. "You know she's half-crazy."

None of the household troubled about Toutouque's private affairs. She was free to come and go, to push the swing door with her nose, to pass the time of day with the butcher or to join my father at his game of écarté; no one feared that Toutouque would stray or that she would get into mischief. So when the restaurant keeper came to inform us, accusing Toutouque, that his bitch Bianca had received a torn ear, we all burst into derisive laughter, pointing to Toutouque, sprawling blissfully, and being clawed by an imperious kitten.

Next morning I had established myself, like a stylite, on the top of one of the pillars connected by the garden railings, and was preaching to an invisible multitude, when I heard approaching a babel of canine howls, dominated by the shrill and desperate voice of Bianca. Then she appeared, dishevelled and haggard, passed the corner of the Rue de la Roche and fled down the Rue des Vignes. At her heels there rolled, with incredible speed, a kind of bristling yellow monster, its legs alternately tucked under its belly and splayed out on all sides, like those of a frog, by

the headlong fury of its advance—a yellow creature with a black mask garnished with teeth, bulging eyes and a purple tongue flecked with foam. It flashed past like a whirlwind and was gone, and while I hastened to descend from my column, I heard from afar the clash, the stormy snarling of a very brief encounter and once again the voice of the red bitch, sorely wounded. I ran across the garden, reached the street door and stood still in amazement; Toutouque, the monster I had glimpsed, yellow and murderous, Toutouque was there, lying at my feet on the steps.

"Toutouque!"

She attempted her kind, fostering smile, but she was gasping and the whites of her eyes, streaked with bloodshot lines, looked as if they were bleeding.

"Toutouque! Is it possible?"

She got up, squirmed heavily and tried to change the subject, but the black lips, the tongue that sought to lick my fingers, still had on them red-gold hairs torn from Bianca.

"Oh! Toutouque! Toutouque!"

I could find no other words in which to express my dismay, my alarm and my astonishment at seeing an evil power, whose very name was unknown to my ten years, so transform the gentlest of creatures into a savage brute.

*Border collie*

*Old English sheep dog*

*Collie*

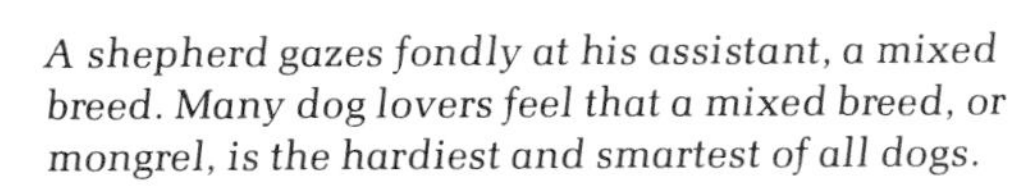

*A shepherd gazes fondly at his assistant, a mixed breed. Many dog lovers feel that a mixed breed, or mongrel, is the hardiest and smartest of all dogs.*

*A kelpie (right) stays with its sheep to the shed door. The sheep trust the familiar dog and put up with its unorthodox exit.*

# Sheep May Safely Graze

One of the earliest duties performed for human beings by dogs was certainly herding sheep. In the first centuries of domestic sheep raising, any sturdy, energetic dog that displayed intelligence and self-reliance no doubt was trained for the task, and puppies born of a good herder were scrutinized for the same abilities. In remote regions of the world, a sheepherding dog is often left alone with its charges for weeks at a time to keep the strays with the herd and drive off any marauding wolves or coyotes. Before modern conveyances, one of the sheep dog's additional responsibilities was helping to take the sheep to market.

The border collie has been herding sheep for farmers in Scotland for four centuries; the Old English sheepdog, despite its name, for a relatively brief 200 years. The venerable rough-coated collie of the Scottish Highlands may date back to Biblical times, and it is thought that the Romans may have brought it to Britain. The Australian kelpie is a recent addition to this hard-working crew of dogs. A mixture of the border collie and the feral dingo, its senses, including its eyesight, are exceptionally developed. The kelpie follows orders that are given from a distance by hand signals and by whistles.

*Malamutes are often teamed in a "gang hitch" (left), harnessed two by two behind a leader. One of the pair below has a forbidding, wolfish look, with dark coloring around the eyes in a masked effect. In the past, the teeth of team dogs were sometimes knocked out to keep the dogs from biting through the harness.*

# Alaskan Sledders

The malamute, or Alaskan sled dog, has played a legendary part in the transportation history of that icebound region. The ancestor of the malamute was probably a Siberian breed introduced into Alaska 2,000 years ago by migrating Eskimo tribes. Although the malamute is often called a husky, only the standardized Siberian is officially referred to by that name.

The malamute's size and bearing, and its almond-shaped eyes and grayish-black coloring resemble those of the wolf, but a genetic connection has never been proved. The animals take their name from the Malamute tribe, which settled on the shores of Kotzebue Sound in northwestern Alaska. In this remote and isolated region, the Malamutes were dependent on the hardy draft dogs, which were kept outside year round. However, the Malamute people generally took good care of the dogs, and in turn the animals were affectionate. As working dogs with a great instinct for survival, they were deliberately bred for the strength needed to maneuver cargoes on rigorous trails across frozen expanses of Arctic tundra.

# Patrician Hounds

Afghan and Saluki are names as elegant and exotic as the hounds they refer to, and the dogs' origins are as obscure as their looks are striking. But intriguing references to both have been found in ancient Sumerian art and in written records from the Upper Nile Valley in Egypt; the Saluki—a royal dog of Egypt—may be the oldest breed of domesticated dog. Its credentials as a superb hunting animal were firmly established by Egyptian pharaohs and Arab sheiks, who tracked gazelles, jackals and foxes with the aid of the keen-sighted hounds: Salukis and Afghans are among the few dogs that hunt by sight instead of scent. Introduced into Britain in the mid-19th Century, the Saluki has earned a reputation as a fast racer and a sleek, popular show dog.

The haughty Afghan was bred selectively by various nomadic desert tribes, but it was life in the rugged hill country of Afghanistan that gave definition to the breed. Inhospitable terrain and seasonal extremes in temperature challenged the Afghan's sturdy constitution, and the dog adapted so well to the mountainous habitat that it became legendary for its surefootedness and agility. Today's Afghan has a natural hunting capability and is an unparalleled hurdle racer.

*The racing form of the Afghan, based on its hunting stride, is impeccable. Its hipbones are high and wide set, allowing maximum flexibility in case the dog must change direction suddenly in pursuit of quarry.*

*The Afghan hound (right) is valued for its aristocratic carriage and regal bearing. Fine, silky hair frames its patrician face. Despite its refined features and the gentle, faraway expression of its eyes, the Afghan can be aggressive.*

*A Saluki (left, bottom) tests its mettle against the sea. Silhouetted against this ocean backdrop, its elegant body contours are clearly visible. The ears and tail are feathered, but the body hair is short. The hindquarters and legs are trim, the hips elevated.*

*Curtains of hair shield the eyes of the Skye terriers at right from the brambles of their rugged native island.*

*A purebred wire-haired fox terrier like the one showing off below is often head and shoulders above other dog-show contenders.*

# Feisty Terriers

In the 19th Century a British parson, the Reverend Jack Russell, developed the white Jack Russell terrier strain, a breed that soon was acclaimed for its intelligence and tenacity: If a fox or badger went to ground, the compact, strong-jawed Russell would unhesitatingly pursue and ferret out its quarry. Such characteristics mark all terriers to some degree, including such older breeds as the Airedale and the fox terrier. The origin of terriers is often traced to the black-and-tan Old English terriers kept by Yorkshiremen to hunt foxes and small game. These dogs were sharp-eyed, determined hunters, and their modern pedigreed counterparts, though differing greatly in appearance from the ancestral stock, retain the same indomitable spirit and keen hunting instincts. They make fine ratters and keep a variety of other pests well under control.

Descended from an independent terrier strain, the irascible, energetic Skye breed has changed little in over 400 years. Natives of Scotland's rugged Isle of Skye, these deceptively harmless-looking small dogs are unsurpassed at controlling vermin and are equally at home running on rocky terrain or swimming across country streams.

*A young Jack Russell terrier (above) dances a canine ballet, and appears quite satisfied with its own impromptu performance.*

*Often called King of the Terriers, the Airedale (left) has an air of regal independence. This breed frequently wins best of show in England and America.*

*Cavalier King Charles spaniel*

*Chihuahua*

*Pekinese*

*Yorkshire terrier*

# Lap Warmers

Of far greater status than stature, the Pekinese and the Chihuahua—at two to six pounds, the world's smallest dog—can be traced to the Eighth Century. The Chihuahua's ancestor was a sacred dog of the Toltecs in Mexico. The revered "Peke" was restricted to the Imperial household in China until 1860, when the British were victorious in a battle over foreign settlement in Peking; while looting the Imperial Palace, they took four Pekes and sent them to England—presenting one to Queen Victoria.

The toy spaniel also may have originated in the Orient, in ancient Japan. It turned up in England in the early 16th Century, where it was called "the Comforter." In this century, the dog has been bred back to resemble more closely its early toy spaniel ancestor. Known as the Cavalier King Charles spaniel, it is the largest of the English toys. The Yorkshire terrier, which appeared for the first time in 19th Century England, was bred down in size from the Skye terrier in only 20 years.

*Toy poodles pose on a park bench (above). Like all toys, they were bred down in size from larger breeds, in this case from the standard poodle in Germany, to serve solely as pets.*

# Sheep, Goats and Pigs

Apart from dogs, goats and sheep have the longest standing as domestic animals, and pigs are not far behind them. Along with cattle and poultry, the three rank highest in their importance to human beings as a primary source of food. During a recent year, one billion sheep, representing more than 800 breeds, yielded almost seven million tons of meat as well as five and a half billion tons of wool. The world's 663 million pigs, classified in some 300 varieties, produced 43 million tons of meat. Goats, numbering 427 million worldwide, are a highly dependable and practical meat and dairy animal. Hardy enough to thrive in debilitating tropical climates or harsh mountainous ones, and easy and inexpensive to care for because they eat virtually any form of vegetation, goats are a mainstay of peoples in nonindustrial countries.

To many people, sheep mean wool rather than mutton or lamb. Yet sheep were first domesticated for their meat and their fat, which served as cooking oil and also, in unforested regions, as fuel. Wool was ignored because wild sheep and early domesticated sheep had coats of long, stiff guard hairs that concealed scanty fleece, if any. The primitive Soay sheep of the Hebrides Islands off Scotland and some sheep in tropical countries where thick, long wool is unnecessary to keep the animal warm still have more hair than fleece. In warm climates, the people had little to fear from chills; in Egypt and other parts of the Near East, sheep still signify meat rather than wool.

In the cold regions of Western Asia, however, where sheep breeding probably originated, the sheep's wool grew long and thick in winter and fell out with the approach of warmer summer temperatures. To people already familiar with turning plant fibers into cloth, the molted wool invited spinning and weaving or, for still-nomadic peoples, the simple process of compressing into felt. As time went on, shepherds began to recognize the logic of cutting the wool off in the spring instead of waiting for it to be shed. After years of being sheared, the sheep adapted to this interference from man and stopped molting.

As wool began proving itself useful, breeders began holding back the ewes and rams with the thickest coats for reproduction and slaughtering the rest of the herd. But the larger animals among those chosen for breeding were often difficult to handle during shearing, so moderate size became another consideration in selection.

Modern breeds of sheep vary considerably from one another largely in superficial ways (such as the quality of their wool), but all of them in general differ substantially from their wild ancestor. Wild sheep have very short tails; many types of domestic sheep have long tails—some so long that they would drag on the ground if they were not clipped. The long tails of some sheep are also very fat and are prized as a delicacy in parts of India and Asia Minor. Such long-tailed sheep are even sometimes equipped with wheeled carts on which to carry their tails—a practice established at least as long ago as the Fifth Century B.C., when the Greek historian Herodotus reported it. Another basic difference between wild and domestic sheep is color: Wild sheep are brown; domestic sheep can be white or black or sometimes varicolored. Wild sheep have large horns, which are greatly diminished in many domestic breeds—in some the ewes have no horns at all. The ears of wild sheep are stiff and small; whereas many domestic sheep have long, floppy ears.

The goat, possibly the first ruminant beast to come under man's control, basically differs very little from its wild ancestor. In fact archeologists have a difficult time determining whether a goat skeleton is that of a wild or domestic animal—a fact that can be determined only by examining the shape of the horn's inner core. Goats were kept by virtually all European and Near Eastern peoples as early as Neolithic times, probably as a source of milk, which is still an important product, especially in the form of cheese. With the notable exceptions of Cashmere and Angora goats, the hair is shorter and less serviceable than sheep's wool, but the hides make good leather.

The pig, almost equal to the steer as a producer of the world's meat, endures its undeserved reputation as a slob with aplomb. Clean as goats in the wild, pigs are far more interesting and versatile domestic animals. In gregariousness and curiosity they resemble dogs more closely than any other domesticated creature. Indeed, from the 11th to the 15th Century in England, pigs were substituted for dogs in the hunt. A decree of William the Conqueror's forbade large dogs in the royal preserve—a measure intended by the King to monopolize hunting. To evade the law, commoners employed pigs as retrievers. As recently as the last century, a sow named Slut reportedly came on the run whenever she was called to the hunt, leaped delightedly at the sight of a gun, and spotted, pointed and retrieved game for her master with the skill of a trained setter.



*Fat-tailed sheep*

*A Heideschnucke ewe (left) tends her young. The mother's creamy fleece contrasts with her dark face and brown-black legs. The Heideschnucke's short tail is characteristic of primitive sheep.*

*The fleece of the Dorset (right) is tightly curled. Sheep's wool is composed of many fibers covered with scales that make the fibers curl and cling to each other, producing long threads of wool desirable for spinning. A Dorset ewe can yield eight to 12 pounds of fleece in a single shearing.*

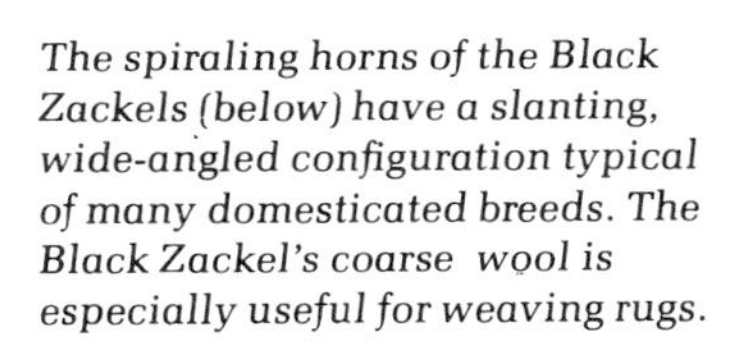

*The spiraling horns of the Black Zackels (below) have a slanting, wide-angled configuration typical of many domesticated breeds. The Black Zackel's coarse wool is especially useful for weaving rugs.*

*Lowland sheep feast in a lush Irish pasture (below). Such sheep are a source of the fine wool that is used in worsteds. Highland sheep supply coarser wool for tweeds.*

# Golden Fleece

From their earliest domestication, sheep have been among the easiest and most rewarding animals to raise. On the Western prairies of the United States flocks of more than 20,000 sheep often forage unattended. But only one acre of good pasture, even if it is too hilly for anything else, can support three or four sheep, and a single sheep can yield about 10 pounds of fleece a year, as well as meat, depending on its variety. A less familiar contribution is the sheep's production of milk. In Spain and Greece, most of the income from sheep is derived from their milk. And in Israel, sheep raisers bring their ewes to "milking parlors," where the fresh milk is bought for making yogurt and cheese. But the major sheep-breeding countries, Australia and the Soviet Union, each with over 140 million head, raise them principally for their wool.

A breed such as the Dorset exemplifies the practicality of sheep raising. Originally from Britain, the all-purpose Dorset matures early, reproduces often, and is a source of quality meat as well as of wool. The Black Zackel is an Eastern European sheep, bred in the Balkans. Its dark wool is particularly thick and shaggy, and it is considered a "woolly" breed. The Heideschnucke, a relatively ancient Northern European breed, produces a white wool, but its varicolored body is characteristic of many similar primitive sheep.

A group of lambs frisks in a Montana meadow (left). These sheep normally breed in the autumn, and after a gestation of five months, the lambs are born in the spring.

*A solicitous ewe (above) has sequestered her newborns in a leafy glade while she licks them clean. The lambs will reach maturity between eight and 10 months of age.*

*A Navajo herdsman drives his flock. The Navajos began herding sheep in the 17th Century; the original stock was brought to the New World by the Spaniards.*

# The Poor Man's Cow

Far from requiring expansive grassy grazing ranges, goats actually prefer to browse on scrub vegetation. They are often raised successfully in arid mountainous regions or on infertile steppes where cattle or sheep could not survive. Goats can be a great resource for rural villagers, providing meat and easily digestible, protein-rich milk. Goatskin is a sturdy hide, and kidskins yield supple leather. Of even greater value are the coats of elite goat relatives: The fine hair of one species is the source of cashmere.

Asia, Europe and Africa account for most of the production of goat's milk. The Nubian and Swiss Alpine are two major breeds. The Nubian originated in Upper Egypt and Ethiopia and is usually black or brown with short, fine hair. The Alpine originated in France and Switzerland, and its short hair, often dappled, ranges in color from white to black. Of the five million goats in the United States, over four million are Angoras, commercially raised for the production of mohair.

*A hornless Nubian billy goat chews a mouthful of brush. The females of such dairy breeds produce highly digestible milk that is often a boon for people whose systems are unable to tolerate cow's milk.*

*Three well-cared-for Alpine Swiss goats (right)—a doe and two offspring—lie in the sun. Swiss herdsmen constantly groom their goats to keep the coats clean and free from parasites.*

A French peasant woman uses a rope leash to restrain her pig (left) as it eagerly hunts truffles. When the pig begins to root around in one particular spot and loosen the soil, the leash is pulled taut and the truffle unearthed (left, bottom).

A sturdy brood sow (below) prepares to ease herself into the mud bath that pigs, which are unable to sweat, wallow in to lower their body temperatures in hot weather. The litter size of sows varies; some breeds can produce more than 300 piglets in a lifetime.

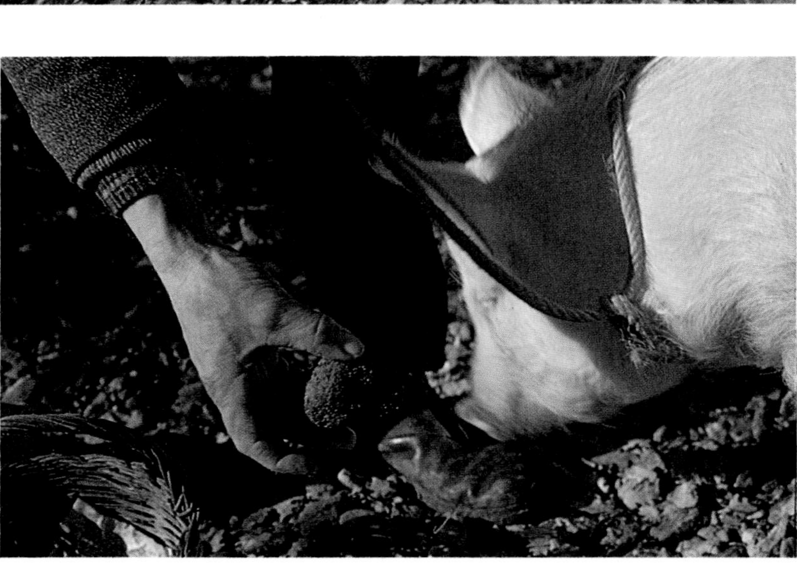

# Productive Porkers

The domestic pig has no peer as a source of protein; it is almost three times as efficient as a steer in converting the same amount of feed into meat and fat. Raised locally and fed on refuse, pigs are the mainstay of the meat supply in China, which raises four times as many pigs as the United States at little cost. But most commercially raised pigs—expensively spoonfed, inoculated and restrained—are a far cry from their free-roaming, scavenging predecessors, even tipping the scales in specialized categories: The carcasses of the heaviest, or "lard" types, weigh in at about 220 pounds, smaller "bacon" types are about 150 pounds, and the smallest, or "pork" types, are 100-pound lightweights.

Apart from supplying staple food themselves, some pigs have found a niche as finders of gourmet food. Exporters of truffles, pungent wild fungi that grow underground in France's Périgord region, depend on the pig's acute sense of smell to locate these flavorful delicacies, which cost about $200 a pound in markets in the United States. Usually the pig is muzzled so that it is unable to enjoy the fruits of its labor as it instinctively roots about. Since truffles grow only in the wild, the pig's contribution to the fanciest cooking is invaluable.

# Birds

Birds have been serving humanity for some 7,000 years, since it is thought that man first began keeping poultry in the Neolithic era. Since then, domesticated birds have functioned as night watchmen, alarm clocks, messengers, oracles, pugilists and racers. They have hunted and fished on command. They have called emperors by name and have sung in concerts for royalty. But most of all, birds are important to man as a source of food.

Poultry is the term generally applied to birds that have been domesticated for their flesh or eggs and includes ducks, geese, turkeys, chickens and pigeons. First to be domesticated were probably the greylag geese that migrated to Egypt from Europe and Asia. The Egyptians designed special buildings in which to incubate goose eggs, which were turned throughout the day by attendants.

The domestic duck, a close relative of the goose, appears to have originated in Mesopotamia and the Far East. The turkey is native to the United States, Mexico and Central America and was the only domesticated animal, other than the dog, that the conquistadors found when they came to the Americas. They soon introduced it to Europe.

The most common species of poultry, however, is the chicken, which is thought to have descended from one or more species of *Gallus*, the jungle fowl indigenous to Southwestern Asia. Until recently, chickens were part of the livestock on almost every farm. But today poultry production is a commercial operation and most chickens spend their lives in a highly mechanized environment in which their diet, temperature and weight are closely supervised. The primary egg producer in the United States is the Leghorn chicken, while the chickens that most often grace the dinner table are the result of a cross between the Plymouth Rock, a fast-growing breed, and the full-breasted Cornish chicken.

Throughout history, chickens have served several purposes besides that of productive farm animal. They provided entertainment for the Greeks, who were great cockfighting fans, and they served as oracles for the Romans: Priests would arrange the letters of the alphabet on the ground and sprinkle feed among the letters; then they would interpret the resulting "spelling" or arrangement of the letters as chickens pecked at the grain. The cock was a symbol of courage for the tribal Gauls of Western Europe and their descendants among the French, who used it as an emblem after the French Revolution. And for ages, the barnyard cock has served unfailingly as the harbinger of dawn.

Like chickens, pigeons have been raised since prehistoric times for food. But they have also been valued for their remarkable homing ability. As recently as World War II pigeons were parachuted to resistance fighters behind German lines, and they returned home with vital dispatches.

The predatory instincts of the raptorial birds have been exploited by people over the ages. Species of falcons have been trained to hunt for their masters since the Assyrians began hawking in the Seventh Century B.C.

Perhaps the most popular domesticated birds are those that serve no other purpose than to entertain. Caged songbirds delighted the Sumerians, the Egyptians and the people of the Indus Valley more than 4,000 years ago. But today's commonest domestic songbirds, the canaries, became popular relatively recently. Native only to the Canary Islands, the Azores and Madeira, they are a drab gray green in the wild and impressed explorers only by their numbers. But individual birds that were brought to Portugal did distinguish themselves vocally. In France, by the time of Louis XIV, a concert by as many as 100 canaries, accompanied by bullfinches and starlings and backed by a three-piece chamber group, was attended even by the King.

Birds that imitate human speech have also long been prized. Mynahs, parrots and parakeets have played a prominent part in holiday festivities in India since ancient times. By 250 B.C., these birds had become popular trade items from India to Alexandria and Rome. The emperor Augustus paid 20,000 sesterces (about $1,500) for a raven that is said to have greeted him each morning with *"Ave Caesar victor imperator"* ("Hail Caesar, victor and emperor").

Budgerigars, widely beloved for their colors as well as their gift of gab, remained unknown outside Australia until the 19th Century. The first "budgie" known to speak was taught in 1789 by Thomas Watling, an inmate at the Port Jackson prison colony in Australia. The colony's physician, James White, visited Watling's hut one day and was welcomed by Watling's pet budgie, which said, "How do you do, Dr. White?" Watling spent his entire life training budgerigars, but it was not until 1842 that a British zoologist on an expedition to Australia discovered Watling's work and publicized the birds' talents. Budgies, in which selective breeding has produced yellow, blue, white and other hues, now surpass even canaries in their numbers and breadth of distribution.

*Domestic goose*

*A flock of geese march single file across a farmyard in Italy. Like their wild counterparts, domestic geese are gregarious, and pairs usually mate for life.*

*An Indonesian farmer (below) oversees a flock of ducks he uses to weed his rice paddy. The birds, which prefer—and therefore eat only—the weeds, are far less harmful to the environment than herbicides.*

# Sacred Geese, Popular Ducks

Waterfowl, with ducks and geese in the lead, were among the earliest animals domesticated by man. Though in ancient Egypt geese were revered as sacred birds, their status was reduced during Roman times to furnishing the feathers used to stuff pillows and mattresses. The Romans also developed a taste for goose livers. Large flocks of the birds were raised just for these organs, which were enlarged artificially by force-feeding, a process that involves cramming a soft mash down the birds' gullets. The technique is still used today, and the resulting livers—some of which may weigh up to two pounds—are used in the delicacy *pâté de foie gras*.

Geese are also raised for their meat, but their popularity worldwide is considerably less than that of ducks, their smaller relatives. The Pekin duck is probably the most prized of all domesticated species. Raised primarily for its meat, the Pekin is a hardy, healthy bird that matures early. Other species of duck are bred for their eggs. The Khaki Campbell duck, for example, surpasses even the chicken (pages 60–61) in fecundity: Some flocks average 320 eggs per bird per year. Ducks also convert feed to flesh with great efficiency, although the high fat content of their meat makes them too rich for the taste of many people, particularly in the Western Hemisphere.

*A Lakenvelder chicken (left) struts across a patch of grass. Popular in medieval times as a show bird, the Lakenvelder is raised today for the same purpose.*

*The shape of the fleshy protuberance on the Rhode Island Red rooster's head (right), called a single comb, is one of its identifying features. This breed is one of the commonest egg and meat producers.*

*Two types of exhibition chicken are the Polish (below, left) and the Bantam (below). Polish chickens, which actually hail from Italy, are named for the Italian word for their topknot of feathers. The name Bantam may be given to miniature versions of many chicken strains.*

# All Cooped Up

Because the chicken is the bird of greatest economic importance to man, it has become the most universal and populous of all domestic bird species. For centuries, most domestic chickens were kept in small flocks and given the run of the barnyard, where they exhibit the same behavior as their wild relatives: One cock presides over a harem of hens, which he defends from intruders. Hens observe a fairly strict hierarchy, with mother hens, which are responsible for providing the chicks with warmth and protection, higher in the pecking order than other hens. Today, most chickens are part of large-scale poultry-farming operations and live in coops where their diet and the temperature and humidity can be regulated.

There are chickens, however, such as the Lakenvelder, Polish and Bantam types shown on these pages, that are bred solely for their ornamental value. This practice also helps preserve a variety of strains, which may be needed by poultry farmers in the future to improve their stock.

The absence of feathers on this turkey's head and neck (left) reveals its red, fleshy crop. The crop, which is an enlargement of the bird's gullet, serves as a receptacle for predigested foods.

# Bigger, Not Better

Turkeys are the only domestic fowl indigenous to the Americas. All the members of the domestic race originated with the Mexican turkey, which Spanish explorers brought back to Spain with them in 1523. During the ensuing decades it appeared in England and the rest of Europe. Today, 115 to 120 million turkeys are raised annually in the United States alone.

Virtually all the turkeys bred for commercial use in the United States are of the broad-breasted variety. The type was developed by breeders in response to consumer preference for large-bodied birds, using only the largest individuals with the broadest breasts for breeding flocks. Such imposed standards go against the pressures of natural selection, which tends to favor medium-sized birds capable of flight. With their size and weakened wing muscles, cultivated turkeys cannot fly. Another unfortunate tendency of broad-breasted birds is a poor reproductive rate, and breeders have been forced to resort to artificial insemination to maintain a sufficient volume to meet the enormous demand.

*Hundreds of turkeys mill around their communal pen on a turkey farm in Oregon. The turkey is the largest of all the domesticated birds, with reported weights of as much as 70 pounds.*

A pair of budgerigars (left) affectionately nibble each other as they sit on the perch in their cage. Natives of Australia, the gregarious birds gather in enormous groups around common waterholes in the wild and should be kept in pairs in captivity.

A shiny red parakeet (right) shows off its sharply contrasting colors. In nature these bold hues actually help conceal the bird in its jungle treetop home by serving to break up its body outline.

Subtle shades of orange and gold mark a quintet of canaries (below). The fine singing of the birds led to their widespread popularity, and young birds are often taught to sing by being exposed to other birds of superior singing ability—and even to musical instruments.

# Caged Vocalists

The domestication of canaries, budgerigars and other parakeets, and parrots is based solely on the pleasure these brightly colored birds bring to their human owners. In addition to the visual appeal of these cage birds, the ability of some to make cheerful noises has also been important in their fascination for human beings. In an odd way, the birds seem to reciprocate. Only those parrots, for example, that are kept in captivity develop the capacity to mimic human speech, whereas large parrots in the wild are loud, screeching birds with a limited range of vocal intonations.

Their association with man has wrought other changes in these bird species. The canary, once the most popular and numerous of all the cage birds, is a nondescript greenish or yellowish-brown color in its native habitat in the Canary Islands, Madeira and the Azores. Centuries of selective breeding by bird fanciers, however, has resulted in a wide variety of hues.

Today the budgerigar is the most popular of the cage birds, which is remarkable considering that, until about 100 years ago, budgerigars were found only in Australia.

*After a training flight, pigeons perch on the roof of a Massachusetts field station run by the University of the State of New York at Stony Brook. Once a bird can "home" from 50 miles, it is used for avian navigation experiments.*

# Expert Navigators

Ever since Noah sent pigeons out from the Ark to investigate whether the flood waters had receded, man has been aware of these birds' remarkable talent for finding their way home over long distances. Roman generals used pigeons to tell the people back home of Caesar's victories in Gaul and, more recently, the birds served as messengers in both World Wars; some were even awarded medals.

Although technology has usurped much of the pigeons' role as couriers, these birds still play a part in science and sport. With their help, scientists have made inroads into the age-old mystery of avian navigation. Recent experiments have shown that pigeons depend in part on the position of the sun and in part on the earth's magnetic field to orient themselves. Just how they do this still remains unexplained, however. Thousands of other pigeons are kept as racing birds and are bred for their speed and endurance. Although the sport has a relatively modest following in the United States, it is extremely popular in Europe. In Belgium, for example, winning birds are often honored with ticker-tape parades.

*A man stoops to duck the wings of the 2,500 racing pigeons he has just released from their cages atop a truck. In this contest, the first bird to finish the 170-mile trip from Lebanon, Pennsylvania, to New York City wins.*

# Cattle

One day in 1521, a Spanish sea captain named Don Gregorio de Villalobos anchored in the sunny Mexican port of what is now Veracruz and sent ashore a cargo of livestock; it included a black bull of the fighting breed and a cow representing the common milch, meat and work animal of the 16th Century. Don Gregorio had brought them from Santo Domingo, where Columbus had landed their great-grandparents in 1493. Thus arrived on the North American mainland the domesticated descendants of the aurochs, the wild ancestor of all modern breeds of cattle. The aurochs, which may have originated in India and begun roaming Europe 250,000 years ago, disappeared in 1627 with the death in a Polish preserve of the last surviving female. Its domestication about 6500 B.C. in the Near East represented one of civilization's most notable advances.

Though the aurochs is extinct, the paintings by Cro-Magnon cave dwellers and a detailed 17th Century drawing of the last aurochs supply an accurate image. Imaginative zoological experiments in this century have disclosed a great deal more not only about the aurochs' looks but about its behavior. Two German scientists, the brothers Lutz and Heinz Heck, have even created reasonable live facsimiles of the animal in a fascinating exercise in genetic research. Through domestication and natural mutation, most modern cattle have lost the ancestral characteristics of the aurochs. A few breeds retain one or more. Through animal generations, the Hecks crossed and recrossed cattle strains having aurochs-like horns with other strains that match the aurochs in coloration or in size.

The Hecks finally developed two reconstituted versions of the aurochs that are thriving and multiplying in European zoos and preserves. These animals are not perfect duplicates of the genuine article, but they come close enough to demonstrate the differences between modern cows and their ancestors: Physically, the new aurochs are large, fast and agile; temperamentally, they are fierce.

Just when the aurochs was put to the plow—its first use by people except for food and leather—has not been established; the momentous event may have occurred in Mesopotamia some 5,000 years ago. In any case, it made a more extensive development of crop-raising agriculture possible. Since then, cattle have had exotic as well as plebeian roles. In ancient Mesopotamia, the likeness of the bull appeared again and again in art and architecture. In Egypt, cattle trod out the grain, and farmers force-fed them to fatten them for slaughter. But the Apis bull—distinguished by such attributes as a white triangle on its brow and a white crescent on its right side—was worshiped as the incarnation of the god Ptah; it lived luxuriously in carpeted chambers, was provided with a select harem and eventually received a ceremonial burial.

In Minoan Crete, the bull was the central figure in a ritual in which acrobatic dancers seized it by the horns and somersaulted onto its back. Even in modern times, cattle are used in ancient ways; for example, their dung is virtually the only fuel in some parts of the world, and cattle-herding peoples such as the Masai of East Africa use it in building their houses.

Selective breeding is almost as old as domestication, but the pace has quickened since British breeders of the mid-18th Century developed the Aberdeen Angus and the Hereford, among others, to yield more and better beef. The Hereford, introduced into the United States in 1817 by Henry Clay—the great Southern Senator—and a partner, has contributed incalculably to the improvement of the Texas longhorn, the tough descendant of the cattle that Villalobos and his fellow Spaniards brought to America. The Aberdeen Angus was imported in 1873 for the same purpose. Both breeds remain important. Climate, however, guides the cattleman's hand: for the hot weather and semiarid range of southeast Texas, breeders at the the King Ranch developed Santa Gertrudis cattle by crossing Brahman, or zebu, bulls from India with shorthorn cows; the Santa Gertrudis thrives on scant forage and does not mind the heat. On the damp Gulf Coast, where the Brahman was originally imported, it has been crossed with local cattle to provide it immunity not only to heat but to ticks and fever.

In recent years, cattlemen, pressed by the rising costs of feed and labor, have been looking far afield for animals that will grow faster and eat less, while yielding beef of high quality. Names hitherto unfamiliar to Americans have been appearing in cattle-market trade journals. Among the names are Simmental, Charolais, Limousin, Normande and Chianina, all well known in Europe, some—such as the Simmental—for centuries. Enthusiasts of these recent imports argue that the animals grow fast, develop long loins, produce lean meat as well as milk, reproduce easily, and have good dispositions. Whether or not these breeds will prove superior economically to the established American breeds is hotly debated.



*Alpine dairy cow*

# Well-bred Bovines

Compared with the rangy wild ancestor first milked by man, the modern dairy cow is a paragon of productive efficiency. Selective breeding by farmers through the years has brought forth a variety of generous milk producers; a purebred cow may yield an average of 2,400 gallons per year. Major breeds in the United States include Holstein, Jersey, Brown Swiss, Ayrshire and Guernsey. Breeding associations and show-ring competition encourage constant improvement of stock. The Holstein, which originated in the Netherlands' Friesland Province, attains the largest size of the dairy cow breeds, weighing an average of 1,500 pounds. Holsteins have been providing high-quality milk for about 2,000 years. The Jersey was developed on the island of Jersey in the English Channel; it is a smaller breed than the Holstein, with the average mature milk cow weighing in at 1,000 pounds.

*The Holstein at left feeds by pinching herbage between its tongue and lower teeth. Cattle, like all ruminants, lack upper incisors and are unable to bite.*

*A Jersey pauses between mouthfuls of forage. Food is fermented in the rumen, a division of the stomach, then the cud is regurgitated and rechewed.*

# Beef on the Hoof

The Texas longhorns that roamed the open ranges of the American Southwest in frontier days were hardy beasts well able to survive long drives across the arid plains to markets. Their ancestors were Moorish stock that Christopher Columbus had brought from Spain to the West Indies and that were later transported to the American continent. After the Civil War, Texas stockmen started a burgeoning beef industry by rounding up and organizing herds from strays. However, by the 1880s these symbols of the early West were viewed as a losing economic gamble: They did not supply enough good-quality beef to justify their need for enormous ranges, and ranchers began to interbreed and supplement them with more docile and productive British breeds, such as the Aberdeen Angus from Scotland.

As enduring in their way as longhorns were the humpbacked Brahmans, well acclimated to India's scorching heat. But their meat matched their endurance for toughness, and they did not enjoy great success as beef cattle in the United States until they were crossed with other breeds; for example, breeders produced the Santa Gertrudis and the Brangus by crossing Brahmans with shorthorn and Angus stock. Another Scottish purebred is the Highland; first developed in the Hebrides, it has been recorded as a breed since the 12th Century.

*The light-colored Brahman bull (left), with its long, slender muzzle and pendulous dewlap, is typical of the humped cattle that are also known as zebu.*

*A young Angus bullock (right) grazes on shoulder-high grass in the heart of Wyoming's cattle country. Angus are a polled, or naturally hornless, breed, and ideally their ebony-black coloring is unmarred by random white patches.*

*Shaggy Scottish Highland cattle (left, bottom) are a distinctive breed, well adapted to a wintry climate and rugged terrain.*

*A Texas longhorn (below) indulges in a dust bath as a relief from biting flies. The longhorn, with its swayback and humped shoulders, is in marked contrast to stocky, squared-off modern cattle breeds.*

*Herefords on a ranch await extra winter feed. Later, lot grain feeding will produce beef with the proportion of fat many Americans favor. This process is wasteful—it takes eight pounds of grain to produce one pound of hamburger.*

*By cattle breeders' standards the Argentinian Hereford at right is a prizewinner. A successful cattle-show contender must be stocky, well proportioned and massive, with deep shoulders, a thick neck and a broad forehead.*

# The Fatted Calf

In Biblical times the slaying of the fatted calf was a traditional ritual of celebration. Today, fattening the calf is a full-fledged industry, involving ranchers and feedlot operators, and utilizing precise scientific methods to produce 13 million tons of beef per year in the United States alone. One of the most successful "easy fattening" beef breeds is the Hereford. Developed in the English county of Hereford, Herefords were first imported to the United States in the early 19th Century and are now, along with the Angus, the major American breed.

The Hereford, or whiteface, matures early and is prized for its steady, systematic conversion of grass and other forage into top-quality beef. With the additional diet available in commercial feedlots before slaughter—a Hereford steer may consume up to 20 pounds of feedlot grain a day for 90 days—the Hereford accumulates expensive fat-marbled pounds of flesh.

*Teams of oxen (above) haul logs out of the Paraguayan Gran Chaco area. The hardy oxen work almost without fatigue, despite adverse conditions—summer flooding, winter droughts, and temperatures that can soar to 115° F.*

*The banteng oxen in the top photograph at right are humped cattle, but the hump is more elongated and less prominent than that of other zebu, or humped species. At right, two bantengs plow a flooded Indonesian rice paddy.*

# Beasts of the Field

Cattle power is a relatively inexpensive energy source that is still successfully tapped in underdeveloped or rugged regions where modern farming equipment and techniques are impractical or unavailable. In Asia and Africa, oxen—the term can mean either true wild oxen or any castrated domestic bull—are invaluable, used not only as draft animals for pulling plowshares and carts, but as mounts and for a limited source of beef.

Herds of domestic banteng cattle are maintained on Java and Bali in Indonesia, where their wild oxen ancestors are now endangered. The Indo-Chinese or Malays probably were the first to see the usefulness inherent in wild bantengs: Though extremely shy and wary, they are undemanding feeders and have great endurance. Domestic banteng are hardy, efficient workers that can survive quite contentedly on the available diet of grass and herbs, tree bark, and swamp and aquatic plants, and can maintain health in densely forested areas as well as on exposed, grassy steppes or mountainous terrain.

# All Creatures Great and Small

## by James Herriot

*In three books, James Herriot has regaled millions of people with tales of the pleasures and pains of being a veterinarian to an array of animals–like the lamb he holds above–in the English countryside. In the excerpt below, from* All Creatures Great and Small, *Herriot describes his most irksome annual task—testing livestock for tuberculosis. Though he entreated the local farmers, Herriot could never get them to round up their herds for inspection before his arrival. Unfortunately for Herriot, if not for his readers, Mr. Kay proved to be no different.*

"I thought you'd have them inside, Mr. Kay," I said apprehensively.

Mr. Kay tapped out his pipe on to his palm, mixed the sodden dottle with a few strands of villainous looking twist and crammed it back into the bowl. "Nay nay," he said puffing appreciatively, "Ah didn't like to put them in on a grand 'ot day like this. We'll drive them up to that little house." He pointed to a tumbledown grey-stone barn at the summit of the long, steeply sloping pasture and blew out a cloud of choking smoke. "Won't take many minutes."

At his last sentence a cold hand clutched at me. I'd heard these dreadful words so many times before. But maybe it would be all right this time. We made our way to the bottom of the field and got behind the heifers.

"Cush, cush!" cried Mr. Kay.

"Cush, cush!" I added encouragingly, slapping my hands against my thighs.

The heifers stopped pulling the grass and regarded us with mild interest, their jaws moving lazily, then in response to further cries they began to meander casually up the hill. We managed to coax them up to the door of the barn but there they stopped. The leader put her head inside for a moment then turned suddenly and made a dash down the hill. The others followed suit immediately and though we danced about and waved our arms they ran past us as if we weren't there. I looked thoughtfully at the young beasts thundering down the slope, their tails high, kicking up their heels like mustangs; they were enjoying this new game.

Down the hill once more and again the slow wheedling up to the door and again the sudden breakaway. This time one of them tried it on her own and as I galloped vainly to and fro trying to turn her the others charged with glee through the gap and down the slope again.

It was a long, steep hill and as I trudged up for the third time with the sun blazing on my back I began to regret being so conscientious about my clothes; in the instructions to the new L.V.I.'s the Ministry had been explicit that they expected us to be properly attired to carry out our duties. I had taken it to heart and rigged myself out in the required uniform but I realised now that a long oilskin coat and wellingtons was not an ideal outfit for the job in hand. The sweat was trickling into my eyes and my shirt was beginning to cling to me.

When, for the third time, I saw the retreating backs careering joyously down the hill, I though it was time to do something about it.

"Just a minute," I called to the farmer, "I'm getting a bit warm."

I took off the coat, rolled it up and placed it on the grass well away from the barn. And as I made a neat pile of my syringe, the box of tuberculin, my calipers, scissors, notebook and pencil, the thought kept intruding that I was being cheated in some way. After all, Ministry work was easy—any practitioner would tell you that. You didn't have to get up in the middle of the night, you had nice set hours and you never really had to exert yourself. In fact it was money for old rope—a pleasant relaxation from the real thing. I wiped my streaming brow and stood for a few seconds panting gently—this just wasn't fair.

We started again and at the fourth visit to the barn I thought we had won because all but one of the heifers strolled casually inside. But that last one just wouldn't have it. We cushed imploringly, waved and even got near enough to poke at its rump but it stood in the entrance regarding the interior with deep suspicion. Then the heads of its mates began to reappear in the doorway and I knew we had lost again; despite my frantic dancing and shouting they wandered out one by one before joining again in their happy downhill dash. This time I found myself galloping down after them in an agony of frustration.

We had another few tries during which the heifers introduced touches of variation by sometimes breaking

away half way up the hill or occasionally trotting round the back of the barn and peeping at us coyly from behind the old stones before frisking to the bottom again.

After the eighth descent I looked appealingly at Mr. Kay who was relighting his pipe calmly and didn't appear to be troubled in any way. My time schedule was in tatters but I don't think he had noticed that we had been going on like this for about forty minutes.

"Look, we're getting nowhere," I said. "I've got a lot of other work waiting for me. Isn't there anything more we can do?"

The farmer stamped down the twist with his thumb, drew deeply and pleasurably a few times then looked at me with mild surprise. "Well, let's see. We could bring dog out but I don't know as he'll be much good. He's nobbut a young 'un."

He sauntered back to the farmhouse and opened a door. A shaggy cur catapulted out, barking in delight, and Mr. Kay brought him over to the field. "Get away by!" he cried gesturing towards the cattle who had resumed their grazing and the dog streaked behind them. I really began to hope as we went up the hill with the hairy little figure darting in, nipping at the heels, but at the barn the rot set in again. I could see the heifers beginning to sense the inexperience of the dog and one of them managed to kick him briskly under the chin as he came in. The little animal yelped and his tail went down. He stood uncertainly, looking at the beasts, advancing on him now, shaking their horns threateningly, then he seemed to come to a decision and slunk away. The young cattle went after him at increasing speed and in a moment I was looking at the extraordinary spectacle of the dog going flat out down the hill with the heifers drumming close behind him. At the foot he disappeared under a gate and we saw him no more.

Something seemed to give way in my head. "Oh God," I yelled, "we're never going to get these damn things tested! I'll just have to leave them. I don't know what the Ministry is going to say but I've had enough!"

The farmer looked at me ruminatively. He seemed to recognise that I was at breaking point. "Aye, it's no good," he said, tapping his pipe out on his heel. "We'll have to get Sam."

"Sam?"

"Aye, Sam Broadbent. Works for me neighbour. He'll get 'em in all right."

"How's he going to do that?"

"Oh, he can imitate a fly."

For a moment my mind reeled. "Did you say imitate a fly?"

"That's right. A warble fly, tha knows. He's a bit slow is t'lad but by gaw he can imitate a fly. I'll go and get him—he's only two fields down the road."

I watched the farmer's retreating back in disbelief then threw myself down on the ground. At any other time I would have enjoyed lying there on the slope with the sun on my face and the grass cool against my sweating back; the air was still and heavy with the fragrance of clover and when I opened my eyes the gentle curve of the valley floor was a vision of peace. But my mind was a turmoil. I had a full day's Ministry work waiting for me and I was an hour behind time already. I could picture the long succession of farmers waiting for me and cursing me heartily. The tension built in me till I could stand it no longer; I jumped to my feet and ran down to the gate at the foot. I could see along the road from there and was relieved to find that Mr. Kay was on his way back.

Just behind him a large, fat man was riding slowly on a very small bicycle, his heels on the pedals, his feet and knees sticking out at right angles. Tufts of greasy black hair stuck out at random from under a kind of skull cap which looked like an old bowler without the brim.

"Sam's come to give us a hand," said Mr. Kay with an air of quiet triumph.

"Good morning," I said and the big man turned slowly and nodded. The eyes in the round, unshaven face were vacant and incurious and I decided that Sam did indeed look a bit slow. I found it difficult to imagine how he could possibly be of any help.

The heifers, standing near by, watched with languid interest as we came through the gate. They had obviously enjoyed every minute of the morning's entertainment and it seemed they were game for a little more fun if we so desired; but it was up to us, of course—they weren't worried either way.

Sam propped his bicycle against the wall and paced solemnly forward. He made a circle of his thumb and forefinger and placed it to his lips. His cheeks worked as though he was getting everything into place then he took a deep breath. And, from nowhere it seemed came a sudden swelling of angry sound, a vicious humming and buzzing which made me look round in alarm for the enraged insect zooming in for the kill.

The effect on the heifers was electric. Their superior air vanished and was replaced by rigid anxiety; then as the noise increased in volume, they turned and charged up the hill. But it wasn't the carefree frolic of before—no tossing heads, waving tails and kicking heels; this time they kept shoulder to shoulder in a frightened block.

Mr. Kay and I, trotting on either side, directed them yet again up to the building where they formed a group, looking nervously around them.

We had to wait for a short while for Sam to arrive. He was clearly a one-pace man and ascended the slope unhurriedly. At the top he paused to regain his breath, fixed the animals with a blank gaze and carefully adjusted his fingers against his mouth. A moment's tense silence then the humming broke out again, even more furious and insistent than before.

The heifers knew when they were beaten. With a chorus of startled bellows they turned and rushed into the building and I crashed the half door behind them; I stood leaning against it unable to believe my troubles were over. Sam joined me and looked into the dark interior. As if to finally establish his mastery he gave a sudden sharp blast, this time without his fingers, and his victims huddled still closer against the far wall.

A few minutes later, after Sam had left us, I was happily clipping and injecting the necks. I looked up at the farmer. "You know, I can still hardly believe what I saw there. It was like magic. That chap has a wonderful gift."

Mr. Kay looked over the half door and I followed his gaze down the grassy slope to the road. Sam was riding away and the strange black headwear was just visible, bobbing along the top of the wall.

"Aye, he can imitate a fly all right. Poor awd lad, it's t'only thing he's good at."

# Horses, Donkeys and Mules

One of the most compelling figures in Greek mythology is the centaur Chiron—a wise and benevolent creature that is half man, half horse, with the head, arms and torso of the man springing from the horse's forequarters in place of its neck and head. Like the winged horse Pegasus, also of Greek mythology, the centaur is godlike; Chiron symbolizes man's urge to superimpose his intelligence and will on the horse's strength, endurance and speed to the point where the two animals are one.

The ideal may have been born in the admiration accorded the first men who learned how to control the movements of the horse—from its back and from a wheeled platform harnessed to the animal. In any case, although the ass and the mule, close equine relatives, have repeatedly proved their value down through the centuries, it is the horse that has captured the human imagination—in war, in peaceful labor, and in sport and recreation.

The mounts of the first horse soldiers, who were probably Asiatic tribesmen, were small, averaging 14 hands, or about four and a half feet tall at the withers—just behind the juncture of neck and body. Used to pull a light one- or two-man chariot or as a mount for an unarmored archer or spearman, most war-horses probably remained small, though varying in weight, until the emergence of the armored knight in the Middle Ages. The development of the longbow and the crossbow, both armor-piercing weapons, led to heavier armor and required the breeding of an animal capable of carrying as much as 420 pounds of warrior, armor and weaponry. The result was the so-called Great Horse, up to 17 hands, or five feet eight inches tall, and thick-legged, the epitome of equine strength.

When firearms ended the age of chivalry—a word that derives from the concept of a war-horse and its rider—cavalry horses were not needed to carry so much weight. Light-boned breeds became fashionable, and enlisted men rode mounts that were bred to combine speed and agility with stamina. The American Indian, one of the most accomplished horse warriors, rode half naked and lightly armed on a horse descended from equine immigrants, the tough cavalry mounts of the Spanish conquistadors.

The bulk and proportions of the knight's charger have been preserved in the great draft horses, some of the most enormous animals ever bred to work. The Shire, which stands 17 hands (over five and a half feet), can pull five tons and was long indispensable in farm work in England, as was the Percheron in Europe. The Shire resembles the Clydesdale, also an English-bred horse, in the feathers of hair that festoon its lower legs; the Clydesdale was famed, in its working days, for its ability when harnessed in large teams to haul heavy loads.

As modern weaponry has eclipsed the horse in war, fossil fuel and the internal-combustion engine have displaced the animal as a power source. Today, the sport of kings, racing on the flat, is the premier field of equine endeavor—at least measured in popularity and cash flow. A champion Thoroughbred racer like Forego, Horse of the Year from 1974 through 1977, won $1,938,957 in purses over a six-year racing career. The excitement of racing attracts more people than major-league baseball, outdrawing the "national pastime" by almost 12 million in 1977. Fox hunting, polo and various other types of competition, from 4-H Club shows to the Olympic Games, are all on the rise.

The horse comes to all these varied activities through selective breeding and through its nature: essentially docile and obedient, adaptable and capable of learning since it responds to rewards and punishments. As an animal that fears predators in the wild, the horse bolts when frightened—sometimes leaving a rider behind—and hastens back to the stable, which represents food and security. It has a keen sense of smell and a distinctive, sloping retina that allows the animal while grazing to shift its focus from one distance to another quickly and efficiently simply by raising and lowering its head, and it can focus simultaneously on objects on the ground and in the distance.

Though the horse has become primarily an athlete and a means of human recreation, the ass, or donkey, and the mule—offspring of a horse and a donkey—remain essentially hardworking animals, particularly in less-developed countries. The domesticated ass, descended from two wild African races, is a strong and patient animal, able to survive on a meager diet of thistles and straw. Bigger than most asses and sometimes bigger than many horses, the mule is calm of temperament, resistant to extremes of heat and cold, strong, and capable of working in terrain that is too rough for the horse. It is equally well adapted to pulling loads or being ridden. Indeed were it not for the mule's reputation for stubbornness—and its principal weakness, a genetic inability to reproduce itself—the animal would come closer to the image of Chiron than any of its equine kin, including the idealized horse.

A group of yearlings (left) frolic at a Florida breeding farm. Thoroughbreds reach their racing peak at the age of three—becoming eligible to compete for the coveted Triple Crown. Secretariat (below) won that title in 1973.

A Thoroughbred (opposite) is coaxed to swim the length of the narrow therapeutic pool at Flag Is Up Farm in California. Swimming is excellent exercise for horses, and it is used to condition Thoroughbreds with leg injuries.

# Breeding for Speed

The Thoroughbred is the fastest, most graceful and most distinguished of equines. All present-day Thoroughbreds are descended from three renowned 17th Century stallions—the Byerley Turk, the Godolphin Arabian, and the Darley Arabian. These founding fathers were all Arabs—a breed that has been maintained with the utmost care since at least 700 B.C. From their Arab ancestors Thoroughbreds inherit stamina, balance and beauty; a small delicate head, a long, arched neck and slim, strong legs.

Breeding Thoroughbreds is as meticulous a process as the name implies. Should a horse prove a poor breeder, its value plummets, and owners often take out fertility insurance to protect themselves against such a possibility. A stallion may refuse to mate with mares of a certain color, and occasionally one will be so unmotivated that he will stand with his head down and his tail between his legs. Such a shy stallion is worthless as a stud, since breeders ban artificial insemination at present. Thoroughbred breeding in the United States has steadily increased, and Thoroughbreds are now raised in every state except Alaska, with more than 23,000 foaled each year.

*Necks outstretched and heads up, three horses round the turn at the 1978 Belmont Stakes (above). The horse on the left wears blinkers to keep it looking ahead and concentrating on the race.*

# Turning for Home

Horse racing probably originated in early Greek and Roman times. Soldiers were the only full-time riders, and warriors would often go into battle with two horses—alternating steeds to avoid tiring them. When not fighting or marching, the soldiers practiced their riding skills by racing—contestants often competing against one another, using as many as four horses per race.

Racing first became an organized sport in the 17th Century when James I of England established public courses in various parts of the country. Most of these races consisted of a series of contests among mature horses of any breed over four to 12 miles of rough countryside. Racehorses were bred for strength and stamina as well as for speed, because they had to be used for transportation, farming and fighting as well as for diversion.

By the mid-19th Century the Thoroughbred had become the predominant racehorse in Europe, the Americas and England. Racecourses became shorter, and Thoroughbreds were bred for speed and precocity—even racing as early as one year of age. The Kentucky Derby, Belmont Stakes and Preakness Stakes—the three races that make up the American Triple Crown—were all established by 1875, and today horse racing is America's leading spectator sport.

*Affirmed, the winner of the 1978 Belmont Stakes, takes the lead over Alydar (right). Jockeys keep their weight forward over the horse's center of gravity to lessen interference with the speed of its natural running gait.*

# Man o' War

## by Walter Farley

*Man o'War was a champion among racehorses, winning 20 of the 21 races he ran in 1919 and 1920 and setting five world records. His performance and style inspired horse lovers all over the world and led Walter Farley to write this fictional account, excerpted from his novel* Man o' War, *of a beautiful but forbidden ride that stableboy Danny Ryan takes on the magnificent horse.*

Man o' War snorted. Danny hushed him again, speaking quietly with his hands, the language both of them knew best of all. He knew he was breaking every rule in the book. If anyone saw him, he would be fired immediately. But it didn't matter now.

Outside, he looked each way, up and down the shed row. Again the big colt snorted, his eyes bright and ears pricked. The night wind swelled his nostrils and fanned his mane and tail.

Danny walked beside him, keeping one hand on the bridle, the other on the colt's neck. "Shh," he kept repeating.

He led Man o' War into the wind, heading for the open gap in the big track. He wasn't going to take any chances of hurting his colt. He wouldn't ride more than a mile at a slow gallop, just enough to remember forever that he had ridden Man o' War!

The stands loomed in the distance, a hovering bulk of steel and concrete and emptiness. Beneath his hands he felt Man o' War begin to quiver. Even without the tumult of a crowd or the music of a band he was becoming excited. It seemed to Danny that Man o' War sensed the quickening of his heart as he stepped on the track rail and mounted him.

"Easy, Red, easy," Danny kept repeating, but there was no easiness in his own body as he let his weight come to rest in the saddle.

Man o' War shifted beneath him, his movements lightning swift and carrying him onto the track. Danny was ready for him. He had carefully watched other riders move with Man o' War in this very same situation.

"Easy, Red, easy," he said again, and although he tried to keep the anxiety from his voice, he knew it was there for the colt to hear. He took up on the reins. Not too tight, he reminded himself. Don't fight him or you're lost. But take hold or he'll get away from you. There, that's better.

He was riding Man o' War! He was moving him down the track, feeling the Herculean strength beneath him and wondering, oh wondering, if he could control it. The world had never looked so beautiful. No other night had ever held such suspense.

"Slow, Red . . . that's it. No hurry now. Just a gallop. No hurry. Slow . . . slow." His hands, too, pleaded with Man o' War. But with every stride the surging power mounted.

He was standing in the stirrup irons as they went past the long, dark stands, and the wind was cold, stinging his face.

"Easy now, Red." He let his weight fall back in the saddle, knowing that if he kept standing in the irons he wouldn't be able to stay on Man o' War. He felt the mighty leap the second his pants touched the leather. It was comparable to nothing he had ever known before. Almost before Man o' War's hoofs struck the packed dirt of the track he leaped again, throwing Danny forward onto his neck.

Danny was scared now, not so much for himself as for Man o' War. It was important that he shouldn't let the colt go all-out. He shortened rein, taking a snug hold on Man o' War's mouth as he had seen others do. Man o' War was finished with fast workouts. He was being let down. He mustn't extend himself.

"Slower, Red," Danny called, and he shortened the reins still more.

Man o' War didn't like the tight hold that pulled his head against his chest. But he was responding to Danny's commands, for his strides shortened going into the first turn.

Even then they were flying, and Danny's excitement grew along with his ever-mounting confidence that he could control Man o' War. He gloried in the tremendous leaps that carried his colt far above the ground with all four legs almost drawn together! And yet with all of Man o' War's speed and strength, he was no wild-eyed monster, grabbing the bit and rushing headlong around the track.

He was intelligent enough to respond to his rider's will, and that was one of the traits that had made him so great.

Danny felt the tremendous pull of Man o' War at the end of the reins; his arms were already beginning to ache from holding him back. That Man o' War would respond to his rider's wishes did not mean that he wasn't constantly asking to be turned loose! It took strong and experienced hands and arms to keep the reins snug and to let Man o' War know what was wanted of him.

To relieve the strain on his arms, Danny finally had to loosen his hold. There was sudden movement beneath him as Man o' War immediately lengthened stride. The track rail became only a blur and Danny's eyes were dimmed by the rush of the wind. All he could see now was the heavy red mane sweeping against his face, stinging his flesh.

Going into the long backstretch he took up rein again, wondering if his strength would last so he could continue to control this powerful horse. His snug hold became a tight one. Pain stabbed muscles that never had been used for such a task before. Then a growing numbness came to his arms and Danny knew this would be followed by a general weakening.

"Easy . . . easy, Red," he said, his voice, too, weakening beneath the mounting strain.

Man o' War's ears flicked back just once, as if perhaps he was listening to him. But the reins slipped still more in Danny's hands and the big colt thundered on!

The iron bit was hard against the bars on his mouth, and Danny no longer had the strength to hold him back. The furlong poles whipped by and Danny realized that Man o' War was now running as he had always wanted to run! His incredible speed mounted as he began digging into the track still more.

Danny kept the hold he had on him, but lowered his own head, pressing it close to Man o' War's neck. This was not the way he had wanted it to be, but there was nothing he could do. The mistake he had made was in taking Man o' War out on the track at all. All he could hope for now was that his colt was strong enough and sound enough to run all-out without injury.

Faster and faster went Man o' War, running for the sheer love of running. Danny couldn't see, couldn't feel anything but the pumping of giant muscles beneath him and

the thumping of his own heart. Whatever might come, he would never in his life forget this ride! He had no trouble keeping his balance. There were no other horses, no crowding, no slamming of riders and their mounts as it would have been in a race. Just himself and Man o' War, and suddenly they were not even of this world!

His hold on the reins slackened still more. There was no rail, no stands, no track . . . nothing but a flying horse whose hoofs barely touched the ground before coming up again. He knew, by the leaning of Man o' War's body to the left, that they were going around the far turn, and he bent in the same direction with his mount.

Straightening out into the homestretch, Danny found his hands being pulled forward still more. He didn't know if they had lost their strength completely or if he had done it willingly. All he realized was that his hold on Man o' War wasn't strong enough to make the big colt shake his head. Man o' War was running free!

There was no roar from the stands to greet him as he came down the stretch like scorching flame. Only the emptiness of the stands and track welcomed the rapid beat of his hoofs. It made no difference to him or to the boy on his back. He streaked through the night, went past the finish pole, and swept toward the first turn again.

Only then did Danny slip back in his saddle and try to take hold of Man o' War. Ignoring the shooting pains in his arms, he pulled back and prayed that his strength would last until he stopped the colt. For a few seconds his only response was a vigorous shaking of Man o' War's head. But the colt, too, was tiring and Danny succeeded in shortening the reins still more. He eased him over closer to the rail to slow him down, and by the time they had completed the turn he had him in a gallop.

Somewhere in the backstretch, he brought him to a stop and straightened in the saddle. He looked around. All was still except for his own heavy breathing. Quickly, he slipped from the saddle to walk Man o' War and watch every step he made. There was no lameness, no misstep. It would take several hours before he could be certain there was no injury. But the outlook was good, and he had the rest of the night before him . . . and all his life to remember.

Together they walked around the track, lost in the darkness.

*A horse and rider (left) soar over a jump called a spread, made up of a series of lined-up obstacles. Midway between takeoff and landing, the rider cannot help the horse clear an obstacle—except by holding completely still.*

*Horses, riders and hounds begin a hunt in Ireland (right). Some of the 200 hound packs in Britain have been maintained since the early 18th Century. Hunting is also popular in the United States, especially in Kentucky and Virginia.*

*Polo players (below) converge on the ball under the eye of a referee. In the 19th Century, polo was played on true ponies (pages 104–105); though today's mounts are large enough to be classed as horses, they are still called ponies.*

# Sports Afield

Narrower in appeal than racing, polo and fox hunting have challenged equestrians for centuries and are growing more popular. Polo is perhaps the oldest goal game in the world, having originated in China and Persia around 600 A.D. In the 16th Century, Persian kings and their courts—including women—played the game, sometimes with as many as 12 balls at a time. In the modern game, four-man teams are mounted on horses that are trained to gallop at the ball and to push opponents off the ball. The aim of the mallet-armed players is to try to knock the ball into the opponent's goal.

Hunting on horseback is as old as riding itself, but hunting the fox with a pack of hounds and no weapons first became popular in the 18th Century. The hunt was once the private recreation of great landowners, who kept their own packs of hounds. Now many packs are maintained by clubs, and participants come from all walks of life.

An outgrowth of the hunt, show jumping is one of the newest of horse-competition events. The course consists of a series of obstacles that vary in number and arrangement. The winning rider is the one with the fewest jumping mistakes, although in a tie, time may be a consideration.

# Lassos and Lumps

The horsemen of the American West have long had a unique sport: the rodeo. "Rodeo" is Spanish for "livestock roundup," and until the 20th Century a cowboy's work and play were so closely connected that rodeo and roundup were virtually synonymous.

Rodeos originated as informal competitions among cowboys to relieve the boredom of cattle driving. They have become strictly regulated events in which two types of horse figure prominently: the bucking bronco and the cowboy's workhorse. There is no such thing as a bucking breed of horse—any unbroken kicker can be used. In both bareback and saddle bronc riding the cowboy must stay on the horse for at least eight seconds. Steer and calf roping entail riding after the animal, roping it and bringing it to the ground. To compete successfully in these events a cowboy must have a horse that will race unafraid after a calf or a fully horned steer, then obey the cowboy's command to stop, and hold its ground while the struggling animal is brought down. In steer wrestling the horse must get alongside the steer so that the cowboy can jump out of the saddle and grab its horns in order to throw his four-legged opponent. Rodeos are often the only live entertainment available in Western towns, and the more than 3,000 contests held each year draw 25 to 30 million people annually.

*A rider prepares to lasso a fleeing calf (left). The cowboy relies on his mount to get him into a good roping position as quickly as possible. Jim McGonigal was a famous calf roper of the 1890s, and to "do a McGonigal" is common rodeo talk for performing well in a roping event.*

*A cowboy and bronco part company. This horse is wearing a "bucking strap," a very tight and annoying belt around its loins that encourages high kicking. Rodeo philosophy favors neither man nor horse: "There was never a horse that couldn't be rode and never a man that couldn't be throwed."*

# Elegant Airs

Among the most intelligent, graceful and highly trained equines are the Austrian Lippizaners and the horses of France's Cadre Noir.

The Spanish Riding School of Vienna, home of the Lippizaners, is the oldest riding school in the world where the classical movements called the "airs above the ground" have been practiced continuously. Only stallions are taught the airs. Young horses are drilled for several years in the basic walk, trot and canter plus more difficult ground moves. The more talented among them are taught such airs as the levade and pesade, lifts on bent haunches; the croupade, a single leap on the hind legs; and the courbette, several croupades in succession. Such difficult movements demonstrate the horses' extraordinary control and grace.

The horses of the Cadre Noir are particularly remarkable because, unlike the Lippizaners, they are not specially bred: Promising horses of any ancestry may be taught the classical airs. The school was founded in the 16th Century at Saumur as a riding academy for young gentlemen. Ruined during the religious wars of the 17th Century, it was reestablished in 1763 as a cavalry school and, except during the French Revolution, has been in operation ever since.

*Mounted officers of France's Cadre Noir perform a synchronized lift (left). The men of the Cadre are dedicated to "the pursuit of calm, impulsion, straightness, lightness, and their application to all equitation."*

*A Lippizaner stallion executes a perfect capriole—a horizontal leap off all four legs (right). Such sophisticated maneuvers are modeled on the natural leaps of the horse in the wild.*

*A group of Lippizaners at the Spanish Riding School's stud farm in Austria (below) amble from pasture to stables. This breed dates back to 1580, the offspring of Arab and Barb stallions and Andalusian mares.*

*A Clydesdale (left) shows off the high spirits that distinguish it from the other four breeds of heavy horse. The Clydesdale owes its popularity to its amiable disposition and handsome looks.*

*Light draft horses draw a feed wagon to a group of waiting sheep (below). The horse is still invaluable when harsh weather or rough terrain makes machines impractical.*

*A snow-flecked Montana draft horse (above) wears various accoutrements of a working harness: bridle, halter, collar and blinkers.*

# The Heavyweights

Though machines do most of the agricultural work today, the strong, heavy pulling, or "draft," horse was an important farm animal until recently. The use of horses as draft animals dates from pre-Roman times, and they were driven like oxen, in yoked pairs, with lines attached to rings through their noses.

All five breeds of draft horses are descended from the so-called Great Horse, which carried the knights of the Middle Ages. The modern draft horse may be over five and a half feet high, and weigh as much as a ton—twice the weight of a Thoroughbred. A single horse can draw as much as five tons, but a pair can pull 18½ tons. Yet these behemoths are probably the most gentle of all the equines.

The Shire horse's great strength is combined with such docility that it can begin working as early as age three. Its legs and feet are covered with long, shaggy hair called "feathers" that are typical of all draft horses except the Suffolk Punch. The stocky Punch is renowned for its longevity —this sturdy horse is often still at work in its twenties. The Scottish Clydesdale is the best known of the five breeds. Its graceful, lively gait makes it extremely popular for parades and shows. The two French breeds, the Percheron and the Breton horse, are smaller than their British relatives. They were used to haul guns in World War I.

# The Working Class

Long before the first horse was tamed, the donkey, a close equine relative, was hard at work for man. Asses, as donkeys are also known, have been used for riding since antiquity, and during Biblical times herds of asses were often bred and kept just for their milk. Today, this versatile animal is used for riding or carrying a pack. The donkey is especially popular in many Middle Eastern countries because it works so well in hot, dry climates.

The mule has served man since prehistoric times. Hybrid offspring of donkeys and horses, mules are unable to reproduce. From their fathers they inherited their long ears, tufted tails, small hooves and loud brays, but their build and impressive strength they owe to their mothers. The popularity of the mule has oscillated. Jews and Christians under Arab rule around 800 A.D. were forbidden by law from riding horses and had to use mules as mounts. In 15th Century Spain, however, the mule was such a popular animal that there was a serious decline in the number of purebred horses raised. It became a capital offense to ride a mule without a special government permit, and the average rider was forced to get a horse—at that time considered a second-rate alternative.

*A mule team pulls a plow for an Amish farmer (above). The size of a plowing team is determined by the width of the area to be plowed. Eight-mule teams such as this were common on the vast prairies of the American Midwest—where even larger 12-mule teams were often worked.*

*The mule's long ears and rounded snout (left) are similar to those of a donkey, but in overall size and strength the mule is comparable to a horse. Mules are the result of a cross between a female horse and a male donkey. The offspring of a stallion and a female donkey, a "hinny," is a smaller, less versatile working animal.*

*A group of Turkish peasants set off for work with their donkeys (right). One beast is used to pull a cart, and several others serve as pack animals and mounts. The donkey's endurance and surefootedness have established its importance as a mode of transport and labor in the Middle East.*

# A Cowboy's Best Friend

When cattle ranching was at its peak after the Civil War, there were no fences to control the thousands of cattle that wandered across the plains, and the cowboy depended upon his horse for everything. Although modern ranches are highly organized, technological ventures, there are still a few where the broken, largely roadless terrain resembles the West of the past, and where horses are still essential to most ranch activities.

Cow-punching crews consist of six or seven hands, each with a string of five or more horses. At dawn each cowboy saddles his horse for the day. Man and animal work together for the next seven hours roping calves, moving the herd, and bringing pregnant cows to the calving pen. Horse and cowboy break at midday, but are back at work in early afternoon, when the crew may be moving camp and gathering wood and water for the next day.

*A quarterhorse and cowboy separate, or "cut," a chosen cow from the herd (right). Cutting horses must be bright and quick to respond to commands in order to cull a single animal from a herd of thousands.*

*A cowboy leads an Appaloosa—a horse with mottled coloring—and a pair of pack mules over a snowy Idaho trail (below). Cold weather ranch work includes periodic rides, which may take days, to check on stock that is wintering on the range.*

# Islands Breeds

By definition, a pony is any horse no taller than 14.2 hands—about four and a half feet—at the withers. Its characteristically diminutive size, heavy coat, and long mane and tail are traits that developed to help ensure its survival in the wild, and enabled it to thrive in rugged climates where food was scarce.

The ancestors of the popular Shetland breed subsisted on a meager diet of heather and vegetation washed up by the sea onto their homes on the Shetland and Orkney islands north of Scotland. The Chincoteague ponies that run wild today on their island off the coast of Virginia are thought to be descendants of Arab horses that swam to shore from a wrecked Spanish ship more than 300 years ago. The ponies have managed to prosper and breed so well on the island's seemingly inhospitable maze of sand and marshes that recently they were in danger of overpopulation. To prevent this an annual Pony Penning Day was established about 50 years ago. Toward the end of July all the ponies are herded together, driven to one end of the island, and made to swim the quarter mile from there to the mainland. Once on shore they are penned, their health is checked, and about half the foals are rounded up for auction. When the auction is over the remaining ponies are herded back to the island to run free for another year.

*A Shetland pony (above) nibbles the crop held by its young rider. Shetlands are very popular as harness horses and pets for children. Shetlands grow to be between 26 and 46 inches high; foals (left) are often less than a foot.*

*A skewbald Chincoteague pony peers from a cover of underbrush (right). These ponies travel in bands. The mares leave the group only to give birth, and return as soon as their foals can walk.*

*Overleaf: A group of range horses and ponies graze and nuzzle each other on the hills of Washington State. In this region, range animals are rounded up once or twice a year to be sold or used. The rest of the time they are left to roam free.*

# Cats

The domestic cat, *Felis catus*, has been a part of the human household for about 3,500 years. The species is the most recent on the cat family tree to have evolved and the only one not threatened with extinction. Thought to descend from the African wildcat *(Felis libyca)*, domestic cats differ from the ancestral species in having shorter legs and a distinctive skull formation and dentition and in breeding more frequently. On the other hand, differences among the roughly 26 varieties—ranging from the affectionate Abyssinian to the sleek and haughty Siamese and the elegant Persian—are relatively slight. In spite of considerable variations in fur color, length and pattern, all domestic cats are approximately the same shape and size. The species presents no such marked contrasts as the Great Dane and the Chihuahua.

Wild ancestry can undoubtedly be given some credit for the remarkable physical and psychological traits of the domestic version. The ability of a cat to land on its feet is legendary, as is its talent for surviving drops of as much as 60 feet. Cats are splendid hunters whose skill in chasing and catching prey is instinctive—though they often must be taught, usually by their mothers, to kill and eat it. One curious attribute of some cats—usually more closely associated with dogs—is an instinct that draws them to their homes or human families over great distances. Though far from common and little understood by scientists, this urge to go home has had astonishing consequences. A cat named Sugar, for instance, spent 14 months and traveled 1,500 miles to find its family, which had moved from California to Oklahoma.

Cats probably first played a domestic role in ancient Egypt. Tomb paintings dating from 1600 B.C. depict obviously domesticated cats in various poses: one eating a fish, another gnawing a bone, another tied to the leg of a chair with a red ribbon. At first, the Egyptians probably merely tolerated cats, since the animals held down rodent popula tions in granaries. But as the priestly overseers of temple granaries saw how useful cats could be, the animals began to be elevated from humble ratters to gods. Bastet, goddess of fertility and good health, was depicted with the head of a cat, and cats were sacred to her and to her followers. Gestures of reverence for cats included mummifying their bodies after death and burying them in mausoleums along with similarly preserved rodents—food for the afterlife. Members of a family whose cat had died mourned the animal's passing by shaving their eyebrows. The law protected cats: Killing one, even accidentally, was an act punishable by death, as was the theft or removal of a cat from Egypt.

Such restrictions slowed the expansion of the cat's range. But as trade developed, the animals spread with the widening commerce to Palestine, Greece and throughout the Roman Empire—largely in their utilitarian role of ratcatcher—and later to the Near East, India and Asia, where they once again attained a kind of revered status.

The spread of Christianity signaled a time of troubles for cats, particularly in Western Europe, where they were associated with heretical and forbidden pagan religions and with evil spirits. The Middle Ages were the bleakest period in the cat's history. As the reputed pets, or "familiars," of witches and the symbol of Satan, cats were hunted and burned—even crucified—by the thousands. Hundreds of years passed before cats regained a measure of their popularity, beginning in the 18th and 19th centuries. Today, cats vie with dogs as house pets. One American family out of five has a cat, adding up to a population in the United States of 45 million cats and furnishing a market for $780 million in commercial cat food each year.

The domestic cat's popularity is really no great mystery. Indeed, cats have long been a source of fascination for human beings. Perhaps it is their air of aloofness, of independence—a purely feline emanation that never lets their human observers forget that all cats, with the notable exception of lions, are basically solitary creatures. That they should occasionally seem to value the company of people seems somehow astonishing—and flattering. Cats also project an enormous sense of dignity and self-esteem. The same aura of dignity even envelops cats on the rare occasions when they make mistakes—there are few more hilarious spectacles than that of a cat recovering its aplomb following a miscalculated leap from table to windowsill. Finally, few young animals can compete with kittens for sheer, beguiling charm.

Yet these traits are only part of the total picture. In some of its aspects, the medieval image of cats as companions of spirits and possessors of superhuman powers reflects a true human reaction to the animals: the respect paid to beings, however acclimated to home and hearth, that are forever beyond man's power to domesticate completely —untamable and wild.



*Black short-haired cat*

# Some Aristocats

The elite of the domestic cat family are the purebreds. Of the 13 million cats born in the United States each year only about 500,000 are such aristocrats, whose price, depending on the breed, ranges from $200 to $1,000 for a sexually mature adult. A cat must meet two criteria to be considered a purebred: Its ancestry must be known for two generations, and its pedigree must be registered with an organization such as the Cat Fanciers Association in the United States.

Cat experts cite roughly nine original breeds of cat: The Siamese, the Birman (not to be confused with the Burmese) and the Persian, the cats of Gambia, Spain, Portugal and the Cape of Good Hope, and the Blue Chartreuse and the Red Tobolsk, both from Malta. By crossing these original groups, however, breeders have been able to produce many new varieties of cat. Some of these were eventually designated pedigreed, a status based on an agreement among the world's cat fanciers as to certain esthetic standards or characteristics. These traits may include body shape and size, or they may relate to such details as the animal's skin, size of ears, fur length and color. Even the cat's temperament may be a factor. Whatever the trait, however, it must distinguish the animal from all other breeds.

*Called the colourpoint in Britain, the Himalayan (left, top) is a breed that has been created by crossing blue or black longhairs with Siamese cats. The Himalayan is a recognized purebred in both Britain and the United States.*

*The Manx's lack of a tail (right) sets it apart from other cats. A true Manx has no hint of tail and is called a rumpie; some Manx—called stumpies—have vestiges of a tail and are less valuable as show animals.*

*A large, broad head, short nose and fine, silky fur are the distinguishing features of the tabby Persian cat (left, bottom). Although still common in Canada and the United States, the term Persian is no longer used in Britain, where the phrase "long-haired cat" has supplanted it.*

*One of the handsomest of the long-haired cats, the Chinchilla (left) has pure-white fur that is tipped with black on the ears, head, tail and back. Other striking traits are the Chinchilla's brick-red nose and its deep-blue or green eyes.*

*Whether they are seal points (below), with dark-brown faces, ears, legs, feet and tails, or tabby points (right), with dark ears, striped legs and ringed tails, ideally all varieties of Siamese cats are medium sized, lithe and long bodied. Both seal points and tabby points have brilliant blue eyes.*

*One of the few purebreds to have originated in the United States rather than in Britain, the brown Burmese (below) has a uniformly dark seal-brown coat. Prizewinning examples of the brown Burmese have penetrating golden-yellow eyes and no trace of white in their fur.*

Short, soft, fine, wavy fur without any guard hairs characterizes the Devon Rex (above), which is also known for its even temperament and quiet voice, making it a fine house pet.

The Abyssinian (left) has ticked fur banded in black or brown. The plush fur of the shorthair (above) is pure white. For genetic reasons, blue-eyed, white-furred shorthairs have a high incidence of deafness.

*A pair of cats engage in mutual grooming (below), a sign of compatibility. In the absence of a friend, however, a cat is quite content to groom itself (bottom).*

*Distracted by a strange noise, a cat pauses in the middle of a thorough grooming session (right), its hind leg poised effortlessly in midair.*

# Well-groomed Contortionists

Cats are meticulous animals that may devote the better part of their day to grooming. Blessed with extremely limber backbones—cats have 30 vertebrae, five more than adult human beings—that can flex and twist as much as 180 degrees, cats are able to assume positions that rival those of the most agile gymnast; they can reach and clean practically every part of their bodies.

The cat's tongue is its primary cleaning aid. Covered with rasplike protuberances called papillae, the tongue acts like a comb as it is drawn over the fur, smoothing out ruffed-up areas and removing specks of dirt and loosened hairs. Parts of its body such as the head, face and ears that the cat is unable reach with its tongue are cleaned instead with its forelimbs, which the cat licks until moist and then uses like a washcloth.

For most cats, grooming—or being groomed—is an experience that transcends hygiene. Since kittens are cared for by their mothers from the moment they are born, touch is the first sensation a cat associates with affection. Even as they mature, cats that are raised in pairs or groups may frequently resort to mutual grooming as a symbol of their friendly relationship.

# The Story of Webster

## by P. G. Wodehouse

*In "The Story of Webster," excerpted below, the English humorist P. G. Wodehouse tells of Lancelot, a carefree bohemian Londoner who, in a letter from his clergyman uncle, is entrusted with the care of Webster, a cat. With Webster's arrival, it immediately becomes clear to Lancelot—though less so to his fiancée Gladys, who is about to embark on a holiday—just who has custody of whom. In the end, however, Wodehouse demonstrates that everyone, even the most dignified cat, has his Achilles heel.*

PS. In establishing Webster in your home, I am actuated by another motive than the simple desire to see to it that my faithful friend and companion is adequately provided for.

From both a moral and an educative standpoint, I am convinced that Webster's society will prove of inestimable value to you. His advent, indeed, I venture to hope, will be a turning-point in your life. Thrown, as you must be, incessantly among loose and immoral Bohemians, you will find in this cat an example of upright conduct which cannot but act as an antidote to the poison cup of temptation which is, no doubt, hourly pressed to your lips.

PPS. Cream only at midday, and fish not more than three times a week.

He was reading these words for the second time, when the front-door bell rang and he found a man on the steps with a hamper. A discreet mew from within revealed its contents, and Lancelot, carrying it into the studio, cut the strings.

"Hi!" he bellowed, going to the door.

"What's up?" shrieked his betrothed from above.

"The cat's come."

"All right. I'll be down in a jiffy."

Lancelot returned to the studio.

"What ho, Webster!" he said cheerily. "How's the boy?"

The cat did not reply. It was sitting with bent head, performing that wash and brush up which a journey by rail renders so necessary.

In order to facilitate these toilet operations, it had raised its left leg and was holding it rigidly in the air. And there flashed into Lancelot's mind an old superstition handed on to him, for what it was worth, by one of the nurses of his infancy. If, this woman had said, you creep up to a cat when its leg is in the air, and give it a pull, then you make a wish and your wish comes true in thirty days.

It was a pretty fancy, and it seemed to Lancelot that the theory might as well be put to the test. He advanced warily, therefore, and was in the act of extending his fingers for the pull, when Webster, lowering the leg, turned and raised his eyes.

He looked at Lancelot. And suddenly with sickening force there came to Lancelot the realisation of the unpardonable liberty he had been about to take.

Until this moment, though the postscript to his uncle's letter should have warned him, Lancelot Mulliner had had no suspicion of what manner of cat this was that he had taken into his home. Now, for the first time, he saw him steadily and saw him whole.

Webster was very large and very black and very composed. He conveyed the impression of being a cat of deep reserves. Descendant of a long line of ecclesiastical ancestors who had conducted their decorous courtships beneath the shadow of cathedrals and on the back walls of bishops' palaces, he had that exquisite poise which one sees in high dignitaries of the Church. His eyes were clear and steady, and seemed to pierce to the very roots of the young man's soul, filling him with a sense of guilt.

Once, long ago, in his hot childhood, Lancelot, spending his summer holidays at the Deanery, had been so far carried away by ginger-beer and original sin as to plug a senior canon in the leg with his air-gun—only to discover, on turning, that a visiting archdeacon had been a spectator of the entire incident from his immediate rear. As he felt then, when meeting the archdeacon's eye, so did he feel now as Webster's gaze played silently upon him.

Webster, it is true, had not actually raised his eyebrows. But this, Lancelot felt, was simply because he hadn't any.

He backed, blushing.

"Sorry!" he muttered.

There was a pause. Webster continued his steady scrutiny. Lancelot edged towards the door.

"Er—excuse me—just a moment . . ." he mumbled. And, sidling from the room, he ran distractedly upstairs.

"I say," said Lancelot.

"Now what?" asked Gladys.

"Have you finished with the mirror?"

"Why?"

"Well, I—er—I thought," said Lancelot, "that I might as well have a shave."

The girl looked at him, astonished.

"Shave? Why, you shaved only the day before yesterday."

"I know. But, all the same . . . I mean to say, it seems only respectful. That cat, I mean."

"What about him?"

"Well, he seems to expect it, somehow. Nothing actually said, don't you know, but you could tell by his manner. I thought a quick shave and perhaps change into my blue serge suit—"

"He's probably thirsty. Why don't you give him some milk?"

"Could one, do you think?" said Lancelot doubtful. "I mean I hardly seem to know him well enough." He paused. "I say, old girl," he went on, with a touch of hesitation.

"Hullo?"

"I know you won't mind my mentioning it, but you've got a few spots of ink on your nose."

"Of course I have. I always have spots of ink on my nose."

"Well . . . you don't think . . . a quick scrub with a bit of pumice-stone . . . I mean to say, you know how important first impressions are . . ."

The girl stared.

"Lancelot Mulliner," she said, "if you think I'm going to skin my nose to the bone just to please a mangy cat—"

"Sh!" cried Lancelot, in agony.

"Here, let me go down and look at him," said Gladys petulantly.

As they re-entered the studio, Webster was gazing with an air of quiet distaste at an illustration from *La Vie Parisienne* which adorned one of the walls. Lancelot tore it down hastily.

Gladys looked at Webster in an unfriendly way.

"So that's the blighter!"

"Sh!"

"If you want to know what I think," said Gladys, "that cat's been living too high. Doing himself a dashed sight too well. You'd better cut his rations down a bit."

In substance, her criticism was not unjustified. Certainly, there was about Webster more than a suspicion of embonpoint. He had that air of portly well-being which we associate with those who dwell in cathedral closes. But Lancelot winced uncomfortably. He had so hoped that Gladys would make a good impression, and here she was, starting right off by saying the tactless thing.

He longed to explain to Webster that it was only her way; that in the Bohemian circles of which she was such an ornament genial chaff of a personal order was accepted and, indeed, relished. But it was too late. The mischief had been done. Webster turned in a pointed manner and withdrew silently behind the chesterfield.

Gladys, all unconscious, was making preparations for departure.

"Well, bung-ho," she said lightly. "See you in three weeks. I suppose you and that cat'll both be out on the tiles the moment my back's turned."

"Please! Please!" moaned Lancelot. "Please!"

He had caught sight of the tip of a black tail protruding from behind the chesterfield. It was twitching slightly, and Lancelot could read it like a book. With a sickening sense of dismay, he knew that Webster had formed a snap judgment of his fiancée and condemned her as frivolous and unworthy. . . .

[*A telegram from a friend, Bernard Worple, indicates that Lancelot is in serious trouble and that only Gladys' return can save him.*]

Arriving in London, her first impulse was to go straight to Lancelot. But a natural feminine curiosity urged her, before doing so, to call upon Bernard Worple and have light thrown on some of the more abstruse passages in the telegram.

Worple, in his capacity of author, may have tended towards obscurity, but, when confining himself to the spoken word, he told a plain story well and clearly. Five minutes of his society enabled Gladys to obtain a firm grasp on the salient facts, and there appeared on her face that grim, tight-lipped expression which is seen only on the faces of fiancées who have come back from a short holiday to discover that their dear one has been straying in their absence from the straight and narrow path.

"Brenda Carberry-Pirbright, eh?" said Gladys, with ominous calm. "I'll give him Brenda Carberry-Pirbright! My gosh, if one can't go off to Antibes for the merest breather without having one's betrothed getting it up his nose and starting to act like a Mormon Elder, it begins to look a pretty tough world for a girl."

"I blame the cat," he said. "Lancelot, to my mind, is more sinned against than sinning. I consider him to be acting under undue influence or duress."

"How like a man!" said Gladys. "Shoving it all off on to an innocent cat!"

"Lancelot says it has a sort of something in its eye."

"Well, when I meet Lancelot," said Gladys, "he'll find that I have a sort of something in my eye." . . .

It was perhaps an hour and a half later that Lancelot, having wrenched himself with difficulty from the lair of the Philistines, sped homeward in a swift taxi. As always after an extended tete-à-tete with Miss Carberry-Pirbright, he felt dazed and bewildered, as if he had been swimming in a sea of glue and had swallowed a good deal of it. All he could think of clearly was that he wanted a drink and that the materials for that drink were in the cupboard behind the chesterfield in his studio.

He paid the cab and charged in with his tongue rattling dryly against his front teeth. And there before him was Gladys Bingley, whom he had supposed far, far away.

"You!" exclaimed Lancelot.

"Yes, me!" said Gladys. . . .

"Lancelot Mulliner," she said, "you have your choice. Me, on the one hand, Brenda Carberry-Pirbright on the other. I offer you a home where you will be able to smoke in bed, spill the ashes on the floor, wear pyjamas and carpet slippers all day and shave only on Sunday mornings. From her what have you to hope? A house in South Kensington—possibly the Brompton Road—probably with her mother living with you. A life that will be one long round of stiff collars and tight shoes, of morning coats and top hats."

Lancelot quivered, but she went on remorselessly.

"You will be at home on alternate Thursdays, and will be expected to hand the cucumber sandwiches. Every day you will air the dog, till you become a confirmed dog-airer. You will dine out in Bayswater and go for the summer to Bournemouth or Dinard. Choose well, Lancelot Mulliner! I will leave you to think it over. But one last word. If by seven-thirty on the dot you have not presented yourself at 6a Garbidge Mews ready to take me out to dinner at the Ham and Beef, I shall know what to think and shall act accordingly."

And brushing the cigarette ashes from her chin, the girl strode haughtily from the room.

"Gladys!" cried Lancelot.

But she had gone.

For some minutes Lancelot Mulliner remained where he was, stunned. Then, insistently, there came to him the recollection that he had not had that drink. He rushed to the cupboard and produced the bottle. He uncorked it, and was pouring out a lavish stream, when a movement on the floor below him attracted his attention.

Webster was standing there, looking up at him. And in his eyes was that familiar expression of quiet rebuke.

"Scarcely what I have been accustomed to at the Deanery," he seemed to be saying.

Lancelot stood paralysed. The feeling of being bound hand and foot, of being caught in a snare from which there was no escape, had become more poignant than ever. The bottle fell from his nerveless fingers and rolled across the floor, spilling its contents in an amber river, but he was too heavy in spirit to notice it. With a gesture such as Job

might have made on discovering a new boil, he crossed to the window and stood looking moodily out.

Then, turning with a sigh, he looked at Webster again—and, looking, stood spell-bound.

The spectacle which he beheld was of a kind to stun a stronger man than Lancelot Mulliner. At first, he shrank from believing his eyes. Then, slowly, came the realisation that what he saw was no mere figment of a disordered imagination. This unbelievable thing was actually happening.

Webster sat crouched upon the floor beside the widening pool of whisky. But it was not horror and disgust that had caused him to crouch. He was crouched because, crouching, he could get nearer to the stuff and obtain crisper action. His tongue was moving in and out like a piston.

And then abruptly, for one fleeting instant, he stopped lapping and glanced up at Lancelot, and across his face there flitted a quick smile—so genial, so intimate, so full of jovial camaraderie, that the young man found himself automatically smiling back, and not only smiling but winking. And in answer to that wink Webster winked too—a whole-hearted, rougish wink that said as plainly as if he had spoken the words:

"How long has this been going on?"

Then with a slight hiccough he turned back to the task of getting his drink before it soaked into the floor.

Into the murky soul of Lancelot Mulliner there poured a sudden flood of sunshine. It was as if a great burden had been lifted from his shoulders. The intolerable obsession of the last two weeks had ceased to oppress him, and he felt a free man. At the eleventh hour the reprieve had come. Webster, that seeming pillar of austere virtue, was one of the boys, after all. Never again would Lancelot quail beneath his eye. He had the goods on him.

Webster, like the stag at eve, had now drunk his fill. He had left the pool of alcohol and was walking round in slow, meditative circles. From time to time he mewed tentatively, as if he were trying to say "British Constitution". His failure to articulate the syllables appeared to tickle him, for at the end of each attempt he would utter a slow, amused chuckle. It was about this moment that he suddenly broke into a rhythmic dance, not unlike the old Saraband.

It was an interesting spectacle, and at any other time Lancelot would have watched it raptly. But now he was busy at his desk, writing a brief note to Miss Carberry-Pirbright, the burden of which was that if she thought he was coming within a mile of her foul house that night or any other night she had vastly underrated the dodging powers of Lancelot Mulliner.

And what of Webster? The Demon Rum now had him in an iron grip. A lifetime of abstinence had rendered him a ready victim to the fatal fluid. He had now reached the stage when geniality gives way to belligerence. The rather foolish smile had gone from his face, and in its stead there lowered a fighting frown. For a few moments he stood on his hind legs, looking about him for a suitable adversary: then, losing all vestiges of self-control, he ran five times round the room at a high rate of speed and, falling foul of a small footstool, attacked it with the utmost ferocity, sparing neither tooth nor claw.

But Lancelot did not see him. Lancelot was not there. Lancelot was out in Bott Street, hailing a cab.

"6a Garbidge Mews, Fulham," said Lancelot to the driver.

# Kitty Litters

Usually solitary, a cat forms close relationships with other cats when it is raising its young—and at few other times. Although a male cat living under the same roof as his mate may help care for kittens, the bulk of the work falls to the female. Toward the end of the gestation period —63 to 65 days—the female seeks just the right spot for delivery.

Though the average litter is four kittens, a blue point Siamese from Australia named Boccaccio Blue Danielle, gave birth to a litter of 13 in 1969. The event was a numerical feat and a logistical one too since cats have only eight teats. Kittens are helpless at birth: They cannot see, they have no teeth, and they have to be stimulated by their mother's licking in order to excrete wastes. By the time kittens are a month old, however, their senses are functioning, and they are ready to begin sampling all that the world has to offer, as shown on the following pages.

*A 10-week-old kitten submits to its mother's ministrations (right, top). At this age kittens spend much time exploring new places, such as the branches of trees (right, bottom).*

*A litter of kittens clamber all over their mother to suckle (below). By the time they are a few days old each kitten has appropriated one teat, nursing on it exclusively.*

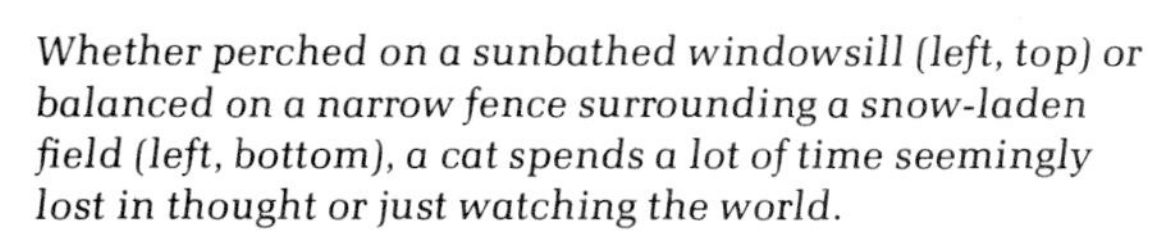

*Whether perched on a sunbathed windowsill (left, top) or balanced on a narrow fence surrounding a snow-laden field (left, bottom), a cat spends a lot of time seemingly lost in thought or just watching the world.*

*The mosaic formed by the slate-gray tiles of a sloping rooftop is a perfect setting for an adventurous and fearlesss marmalade-colored cat catching the warm rays of the morning sun.*

# The Cat and the Moon

### by William Butler Yeats

*Every cat owner has been aware, at one time or inother, of the quality of otherworldliness his inimal exudes, feeling that even the tamest and riendliest domesticated feline has insights into he secrets of the universe that no mere human an hope to share. The Irish poet William Butler Yeats captured this quality in "The Cat and the Moon," which tells of a beautiful interlude between these two "creatures" of the night.*

The cat went here and there
And the moon spun round like a top,
And the nearest kin of the moon,
The creeping cat, looked up.
Black Minnaloushe stared at the moon,
For, wander and wail as he would,
The pure cold light in the sky
Troubled his animal blood.
Minnaloushe runs in the grass
Lifting his delicate feet.
Do you dance, Minnaloushe, do you dance?
When two close kindred meet,
What better than call a dance?
Maybe the moon may learn,
Tired of that courtly fashion,
A new dance turn.
Minnaloushe creeps through the grass
From moonlit place to place,
The sacred moon overhead
Has taken a new phase.
Does Minnaloushe know that his pupils
Will pass from change to change,
And that from round to crescent,
From crescent to round they range?
Minnaloushe creeps through the grass
Alone, important and wise,
And lifts to the changing moon
His changing eyes.

# Bibliography

NOTE: An asterisk at the left means that a paperback volume is also available.

Alcock, Anne, *The Love of Horses*. Octopus Books (London), 1973.

American Kennel Club, *The Complete Dog Book*. Howell Book House, 1972.

Austin, Jr., Oliver L., *Birds of the World*. Ed. by Herbert S. Zim. Golden Press, 1961.

Bastyai, Lorant de, *The Sport of Falconry*. Pelham Books and Charles T. Branford Co., 1968.

Beadle, Muriel, *The Cat: History, Biology, and Behavior*. Simon and Schuster, 1977.

Brereton, J. M., *The Horse in War*. Arco Publishing Co., 1976.

Brown, Leslie, and Dean Amadon, *Eagles, Hawks and Falcons of the World*. McGraw-Hill Book Co., 1968.

Churchill, Peter, *The Sporting Horse*. Arco Publishing Co., 1976.

Cole, H. H., and Magnar Ronning, eds., *Animal Agriculture: The Biology of Domestic Animals and Their Use by Man*. W. H. Freeman, 1974.

Davis, Henry P., ed., *The Modern Dog Encyclopedia*. Stackpole and Heck, 1949.

Dembeck, Hermann, *Animals and Men*. Trans. by Richard and Clara Winston. The Natural History Press, 1965.

Dobie, J. Frank, *The Longhorns*. Little, Brown, 1941.

Epstein, H.:
*Domestic Animals of China*. Africana Publishing Corp., 1967.
*The Origin of the Domestic Animals of Africa*. Africana Publishing Corp., 1971.

*Fletcher, Walter R., *Dogs of the World*. Ridge Press, 1977.

Fox, Michael W.:
*Behaviour of Wolves, Dogs and Related Canids*. Harper & Row, 1971.
*The Dog: Its Domestication and Behavior*. Garland STPM Press, 1978.
**Understanding Your Cat*. Coward, McCann & Geoghegan, 1974.
**Understanding Your Dog*. Coward, McCann & Geoghegan, 1974.

Gianoli, Luigi, *Horses and Horsemanship through the Ages*. Crown Publishers, 1967.

Gooders, John, *The Great Book of Birds*. The Dial Press, 1975.

Hafez, E. S. E., ed., *The Behaviour of Domestic Animals*. Baillière, Tindall & Cox (London), 1962.

*Harris, Marvin, *Cannibals and Kings: The Origins of Cultures*. Vintage Books, 1977.

Henderson, G. N., and D. J. Coffey, eds. *The International Encyclopedia of Cats*. McGraw-Hill Book Co., 1973.

Hope, C. E. G., and G. N. Jackson, eds., *The Encyclopedia of the Horse*. The Viking Press, 1973.

Howey, M. Oldfield, *The Horse in Magic and Myth*. Castle Books, 1958.

Hyams, Edward, *Animals in the Service of Man*. J. B. Lippincott, 1972.

Johnson, Patricia H., *Meet the Horse*. Grosset & Dunlap, 1967.

Levinson, Boris M., *Pets and Human Development*. Charles C. Thomas, 1972.

Mery, Fernand:
*The Life, History and Magic of the Cat*. Trans. by Emma Street. Grosset & Dunlap, 1966.
*The Life, History and Magic of the Dog*. Grosset & Dunlap, 1968.

Pond, Grace, ed., *The Complete Cat Encyclopedia*. Crown Publishers, 1972.

Trippett, Frank, and the Editors of TIME-LIFE BOOKS, *The First Horsemen*. TIME-LIFE BOOKS, 1974.

Vernam, Glenn R., *Man on Horseback*. University of Nebraska Press, 1964.

Walker, Ernest P., *Mammals of the World*, 3rd ed. Vols. 1 and 2. The Johns Hopkins University Press, 1975.

Wood, Gerald L., *The Guinness Book of Animal Facts and Feats*. Guinness Superlatives (Enfield, England), 1972.

Zeuner, Frederick E., *A History of Domesticated Animals*. Harper & Row, 1963.

# Credits

Cover—M. and B. Reed, Animals Animals. 1—W. Hubbell, Woodfin Camp. 5—Peter B. Kaplan. 6–7—K. van den Berg, Photo Researchers, Inc. 10—P. Koch, P.R., Inc. 19—Stephen Green-Armytage. 20—Anthony Edgeworth. 21 (top)—P. Miller, P.R., Inc.; (bottom)—R. Kinne, P.R., Inc. 22 (top)—W. Hubbell, Woodfin Camp; (bottom)—R. Kinne, P.R., Inc. 23—J. Niver, Animals Animals. 24 (top)—S. Thompson, Animal Photography, Ltd.; (bottom)—R. Kinne, P.R., Inc. 25—Nina Leen. 26–27—Stephen Green-Armytage. 27—G. Zahm, Bruce Coleman, Inc. 28 (top)—R. Kinne, P.R., Inc.; (bottom)—L. Georgia, P.R., Inc.; 28–29—J. Dominis, Time Inc.; 29 (top)—R. Kinne, P.R., Inc. 30 (top)—S. Thompson, A.P., Ltd.; 30–31—F. Roche, Animals Animals; 31 (top)—S. Thompson, A.P., Ltd.; (bottom)—J. Cooke, Animals Animals. 36 (top)—Peter B. Kaplan; (center)—Stephen Green-Armytage; (bottom)—F. Roche, Animals Animals; (right)—H. Reinhard, B.C., Inc., 37—J. Burt, B.C., Inc. 38–39—B. Ruth, B.C., Inc.; 39—K. Gunnar, B.C., Inc. 40 (top)—J. Dominis, Time Inc.; (bottom)—Stephen Green-Armytage. 41—M. Bruhmuller, B.C., Inc. 42 (left), 42–43—L. McCombe, Time Inc.; 43 (left)—R. Kinne, P.R., Inc.; (right)—Stephen Green-Armytage. 44 (top left)—S. Thompson, A.P., Ltd., (top right)—J. Cooke, P.R., Inc.; (bottom left)—M. Reed, Animals Animals; (bottom right)—Stephen Green-Armytage, P.R., Inc. 45—F. Grehan, P.R., Inc. 47—V. Englebert, P.R., Inc. 48 (top)—H. Reinhard, B.C., Inc.; (center)—R. Jaques, P.R., Inc.; (bottom)—R. Kinne, P.R., Inc.; 48–49—G. Johnson, P.R., Inc. 50 (top)—N. deVore III, B.C., Inc.; (bottom)—B. Belknap, P.R., Inc.; 50–51—L. Rhodes, Animals Animals. 52—K. Brate, P.R., Inc. 53—K. van den Berg, P.R., Inc. 54–55—A. Woolfitt, Woodfin Camp. 57—T. Myers, P.R., Inc. 58–59 (top)—J. Ross, P.R., Inc.; (bottom)—J. Fields, P.R., Inc. 60 (top)—R. Kinne, P.R., Inc.; (bottom)—Z. Leszczynski, Animals Animals. 61 (top)—T. Myers, P.R., Inc.; (bottom)—D. Specker, Animals Animals. 62—J. Latta, P.R., Inc.; 62–63—M. Vignes, P.R., Inc. 64—H. Reinhard, B.C., Inc. 64–65—H. Reinhard, B.C., Inc.; 65 (top)—K. Fink, B.C., Inc. 66—L. Snyder, 66–67—L. Stewart, SPORTS ILLUSTRATED, Time Inc. 69—P. Miller, P.R., Inc. 70—E. Harris, P.R., Inc. 71—Z. Leszczynski, Animals Animals. 72 (top)—J. and D. Bartlett, B.C., Inc.; (bottom)—H. Silvester, P.R., Inc. 73 (top)—J. Flannery, B.C., Inc.; (bottom)—T. McHugh, P.R., Inc. 74–75—J. Wright, B.C., Inc.; 75—J. Cooke, Animals Animals. 76—L. McCombe, Time Inc. 77 (top)—T. McHugh, P.R., Inc.; (bottom)—G. Holton, P.R., Inc. 78—Derek Bayes. 79—Terence Spencer. 80—Terence Spencer. 83—E. Weiland, P.R., Inc. 84–85 (top)—Stephen Green-Armytage; (bottom)—G. Long, SPORTS ILLUSTRATED, Time Inc.; 85—C. Palek, Animals Animals, 86–87—H. Kluetmeier, SPORTS ILLUSTRATED, Time Inc. 92 (top)—E. Weiland, P.R., Inc.; 92–93—F. Grehan, P.R., Inc.; 93 (top)—E. Weiland, P.R., Inc. 94–95—Wright,

B.C., Inc. 96—E. Weiland, P.R., Inc. 97 (top)—J. Cooke, P.R., Inc.; (bottom)—P. Koch, P.R., Inc. 98 (top)—Stephen Green-Armytage; (bottom)—J. Wright, B.C., Inc.; 98–99—J. Wright, B.C., Inc. 100 (bottom)—K. Fink, B.C., Inc.; 100–101—L. Rhodes, Animals Animals; 101—B. Ray, Time Inc. 102–103—N. deVore III, B.C., Inc.; 103—J. Munroe, P.R., Inc. 104–105 (top)—A. Woolfitt, Woodfin Camp; 105 (bottom)—R. Redden, Animals Animals. 106–107—E. Roberge, P.R., Inc. 109—J. Burton, B.C., Inc. 110 (top)—J. Howard, P.R., Inc.; (bottom)—Stephen Green-Armytage, P.R., Inc. 111—S. Thompson, A.P., Ltd. 112–113—S. Thompson, A.P., Ltd. 114 (top)—Stephen Green-Armytage, P.R., Inc.; (bottom)—T. Dickinson, P.R., Inc. 115—W. Morgan, Animals Animals. 120–121—Peter B. Kaplan. 122 (top)—S. Shea, P.R., Inc.; (bottom)—M. and B. Reed, Animals Animals; 122–123—J. Cooke, Animals Animals.

The photographs on the endpapers are used courtesy of the Time-Life Picture Agency, Russ Kinne and Stephen Dalton of Photo Researchers, Inc., and Nina Leen.

The film sequence on page 8 is from "Chincoteague Ponies," a program in the Time-Life Television series *Wild, Wild World of Animals.*

THE ILLUSTRATIONS: 9—British Museum. 12–13—Enid Kotschnig. 16–17—Metropolitan Museum of Art. 32–35—Gerry Gersten. 89, 90–91—John Groth. 117, 119—Chas. B. Slackman. 124–125—Catherine Siracusa.

# Index